Arithmetic
for Parents

A Book for Grown-Ups About
Children's Mathematics

Revised
Edition

Arithmetic
for Parents

A Book for Grown-Ups About
Children's Mathematics

Revised
Edition

$1+1=2$

$9 \times 6 = 54$

$7+7 = 14$

$5-2=3$

$9 = 17$

Ron Aharoni
Technion, Haifa, Israel

World Scientific

NEW JERSEY · LONDON · SINGAPORE · BEIJING · SHANGHAI · HONG KONG · TAIPEI · CHENNAI

Published by

World Scientific Publishing Co. Pte. Ltd.

5 Toh Tuck Link, Singapore 596224

USA office: 27 Warren Street, Suite 401-402, Hackensack, NJ 07601

UK office: 57 Shelton Street, Covent Garden, London WC2H 9HE

Library of Congress Cataloging-in-Publication Data
Aharoni, Ron.
    Arithmetic for parents : a book for grown-ups about children's mathematics / Ron Aharoni, Technion, Haifa, Israel. -- [2015 edition].
        pages cm
    Translated from Hebrew.
    Previously published: El Cerrito, Calif. : Sumizdat, 2006.
    ISBN 978-9814602891 (hardcover : alk. paper) -- ISBN 978-9814602907 (pbk. : alk. paper)
    1. Mathematics--Study and teaching (Elementary) 2. Arithmetic--Foundations. I. Title.
    QA135.6.A37 2015
    372.7'2--dc23

                                            2014047935

British Library Cataloguing-in-Publication Data
A catalogue record for this book is available from the British Library.

Typeset by Stallion Press
Email: enquiries@stallionpress.com

Printed in Singapore

Contents

Preface

Most adults have long buried their memories of studying mathematics. All they really want is to forget the trauma. They accept their past incomprehension as a tolerable, albeit painful, fact. "You don't really need to know mathematics," they console themselves. Until one day the need does arise and the old anxieties resurface. This happens when their child begins dealing with the same experiences.

Many people would like to help their children with their arithmetic studies, but are too afraid to return to a subject that was so painful for them as children. However, they forget that they have acquired many new tools since their schooldays. The adult has more forbearance, abstraction skills, the capability to deal with complex sentences, and the patience to wait until the whole picture emerges. All these skills can be used to deal more easily and quickly with the principles of elementary school mathematics.

The purpose of this book is to provide the guidance required to do so. It offers assistance to the parent who wishes to be an active participant in his or her child's arithmetic studies. As a matter of fact, that is how the book was born. The parents at my son's school asked for written instructions so that they could help their children with mathematics. What began as modest notes slowly evolved and eventually reached the structure and proportions of a book.

Books, like ideas, have a life of their own. Sometimes they lead their writer no less than he leads them. And so it was that this book gradually took on a different form. One of the insights I came by while teaching in elementary school is that elementary mathematics isn't simple at all. Besides beauty, it has depth. This message slowly found its way into the book and gave it an additional direction: A description of the beauty of elementary mathematics and, consequently, mathematics in general. Thus my original target audience expanded to include the reader who wishes to return to his childhood mathematics, but from a different angle. For this group of readers the book provides a second chance. Those who have learned how to multiply fractions or how to perform long division, but never understood why it was done exactly so, are invited to take a look from a new, mature viewpoint.

The book is also aimed at a third audience, one no less important for me: teachers and educators. To them the book sends a clear message: Proper teaching of mathematics depends more on an understanding of the mathematical principles than on educational tricks. It requires familiarity with the way the fine mathematical layers lay one upon the other. Furthermore, it is best done by direct, concrete experience with the concepts, with no intermediaries.

The first part of the book, **Elements**, describes the fundamentals of elementary mathematics from the point of view of a teacher. That is, it deals with the questions of what mathematics is, what material should be taught in elementary school, and what makes mathematics beautiful and similar to the arts.

The parent who wishes to help his or her child must, beyond mathematics, be familiar with basic teaching principles, and to these I have dedicated the second part of this book. It includes the principal rules of progressing from the concrete to the abstract. In addition, parents must also be familiar with the educational trends that determine the educational character of their child's school. For this reason, the book includes an appendix describing, in a nutshell, the major developments in mathematical education over the last fifty years.

The third part of the book is devoted to the nitty-gritty of elementary mathematics, the details of the material taught in school, stage by stage. Although geometry constitutes between 10% and 20% of the usual curriculum, I decided to renounce it and stick to arithmetic, that is, the study of the properties of numbers. Besides the centrality of arithmetic in the curriculum, the reason for this choice is that arithmetic is a uniform, polished body of knowledge, not unlike a fine diamond, and worthy of a book dedicated to it alone.

First Lesson

You never get a second chance to make a good first impression.

An American saying

In education as in life, first impressions are important. The way a subject is introduced for the first time will determine to a great extent the student's future attitude towards it. Will it remain a pleasant memory, or a painful one? Will it arouse a feeling of "I understand," or of "this is difficult"?

For this reason, chapters titled *First Lesson* are interspersed throughout the book. They provide suggestions of possible ways to present subjects for the first time. There is never a single method, but it is always helpful to have a few options. The ideas are presented as pointers for teachers, but parents may also find them useful.

Introduction

What I Learned in Elementary School

A poet must have his childhood close at hand.

Theodore Roethke, Poet

A friend of mine left the hi-tech industry and a management career in mid-life and decided that his purpose in life is education. Not just any form of education, but mathematical education. In September 2000, just before the school year began, he called and told me of a project to promote mathematical education in Maalot, and that I should join in. (Maalot is a development town in the far north of Israel. These towns were built during the 1950s to settle new immigrants and are usually considered to be rather backward.)

I teach mathematics at the university. Indeed, I have always been interested in teaching, and was involved in youth activities for many years. Among other projects, I used to instruct gifted elementary and high school students. However, I had not set foot in an elementary school since graduating the sixth grade myself. Therefore I consulted whomever I could. The advice I got was more or less unanimous: You have no idea what you're in for. Teaching gifted children is completely different from teaching ordinary children. Teaching in an elementary school is a profession. It is foolish to believe you can rely on your knowledge of the principles of teaching at the university. (At that time I believed that those, at least, I knew.)

I also discussed my dilemma with an experienced teacher whose opinion I valued. Upon hearing the idea, she burst out in a rage I never imagined her capable of. Don't you dare, she yelled at me. People like you are ruining elementary education. You will be no different from all the other academicians who have no idea how to teach in elementary school, and come to coach teachers with their fantasies, wreaking havoc in education. You'll go to Maalot, confuse everyone, and after they've been burned they'll be wary even of cold water for years.

Looking back, I can't believe I agreed to teach in Maalot in spite of all that. With the innocent conceit of a university professor, I assumed that I knew better than those who were, after all, only teachers. Looking back, I realize that heeding the advice I received would have cost me one of the most fascinating adventures of my life.

The slogan I embroidered on my banner was "hands-on experience." I would let the children experiment with mathematical concepts, and after

1

experiencing them concretely, they would easily perform the abstraction. I started teaching in advanced classes — fourth and fifth grades. I took the children into the playground to measure the length of the shadows of trees, posts and buildings. We calculated the ratio between the children's shadows and heights, and used this information to calculate the height of the trees according to their shadows. (This idea is borrowed from Thales, born in the 7th century B.C., the first mathematician in history to be mentioned by name, who used this method to calculate the height of the pyramids.) We drew circles on the pavement, measured radii, diameters and circumferences, and compared them. We measured the length and width of the classrooms in various ways. We learned how many floor tiles fit (lengthwise) in one meter, by computing the ratio between the length of the classroom in tiles and its length in meters.

I learned the price of conceit the hard way. There was very little meaningful teaching going on. Most lessons were a mess.

I remember my first day of insight well. I took a fourth-grade class out to the playground, to measure the diameters and circumferences of circles drawn on the pavement. The teacher of the class was watching patiently the inevitable chaos that resulted as the children took the opportunity to fool around. Eventually she suggested that we return to the classroom. In class we drew circles on the blackboard, and discussed the ratio between circumference and diameter. I was surprised to discover how easy it was to conduct an intelligent discussion with the children. I realized that I was underestimating the children's power of abstraction. I also realized the power of words, and of interactive discussion.

Fortunately, I also began teaching first-graders at that time. This was a thrilling experience. Meeting first-graders is enlightening. They have not yet been corrupted; they trust you and will go along with you wherever you lead them. They respond directly and immediately communicate what works for them and what doesn't. There is no better place to learn how to teach than in the first grade. I also met an excellent teacher there, Marcel Granot, who was willing to join me in an adventure that enriched us both. I would open the lesson, and Marcel would intervene whenever she felt that the didactic aspects were less than perfect. This usually happened when I didn't pace the lesson, that is, when I skipped a step.

From that day on, I have been learning, intensively and constantly, from each lesson and from every teacher I meet. The less successful lessons teach me as much as the successful ones. And the ones I learn the most from are those that start out with a limp and then take off when the right thing is done.

What have I learned? Much about teaching, approaching children, and about the way children think. I have learned about the importance of being systematic, something I desperately lacked at first. I understood that concepts adults perceive as a whole are actually built of many small elements, one upon the other, and that you cannot skip any one of them. I learned from personal experience that explanations are usually futile in elementary school: Concepts must originate in the child through personal experience. I was right about hands-on experience from the start. The problem was that I had no idea how to integrate it into the learning process. Hands-on experience need not necessarily apply to complex notions. It is essential also in acquiring the most basic concepts, such as the concept of the number, or what it means to be "greater than" or "less than."

But aside from all that, I was in for a real surprise. Had I been told that in revisiting the elementary school I would myself learn mathematics, I would have never believed it. To my surprise, that is exactly what happened: I learned a lot of mathematics, perhaps even mainly mathematics. Had I gone to teach in high school, this would probably not have been the case. The professional mathematician is familiar with the mathematics studied there, but in elementary school he is in for some novelties. It is there that the most basic elements arise: the concept of the number and the meaning of arithmetical operations. These are elements that the professional mathematician rarely pauses to consider.

A large part of what I learned wasn't new facts, but something completely different — subtleties. It was like looking at a piece of cloth. From afar it seems smooth and uniform, but up close you discover that it is made of fine, interwoven threads. What I believed to be one piece turned out to consist of a delicate texture of ideas. More importantly, I realized that to be a good teacher one must be familiar with the fine elements and the order by which they are interwoven. "Pacing, pacing, pacing," as Marcel used to remind me.

This book is, to a great extent, about the subtleties that are at the base of mathematics, the subtleties that make it beautiful and provide meaning to its teaching.

Part 1

Elements

What is Mathematics?

The book of nature is written in the language of mathematics.

Johannes Kepler, Astronomer

The Queen of Sciences

Mathematics is the queen of sciences and arithmetic the queen of mathematics.

Carl Friedrich Gauss, Mathematician

In second-grade classes, I try to explain the importance of numbers. I tell the children a story of a king who hated numbers so much that he forbade their use throughout his kingdom. Together, we try to imagine a world free of numbers, and discover that life in it is very limited. Since it is forbidden to mention a child's age, children of all ages enter the first grade. You cannot pay for your groceries, nor can you set up an appointment, since you are not allowed to mention the number of hours and minutes.

This is only an illustration of the importance of mathematics in our lives. As civilization and technology advance, our lives become more and more dependent on mathematics. Steven Weinberg, a Nobel laureate in physics, dedicated two chapters of his book *Dreams of a Final Theory* to subjects beyond physics: mathematics and philosophy. He writes that time and again he is surprised to discover how useful mathematics is, and how futile is philosophy.

To understand why this is so, one must understand what is mathematics. This is not a simple question. Even professional mathematicians find it difficult to answer. Bertrand Russell said of mathematicians that they "don't know what they are doing." (His judgment of philosophers was even harsher: A philosopher in his eyes is "a blind man in a dark room looking for a black cat that isn't there.") This is true in at least one sense: Most mathematicians do not bother to ask themselves what it is, exactly, that they are doing.

To try answering this question, we will start with a simple example: What is the meaning of $3 + 2 = 5$?

In the first grade, I ask the children to examine how many pencils there are when you add 3 pencils to 2 pencils. They know that "addition" means "joining." Therefore they join 3 pencils and 2 pencils and count: 5 pencils. Now I ask, "How many buttons are there when you add 3 buttons and

2 buttons?" "5 buttons," they answer, without missing a beat. "How do you know?" I insist. "We know from the previous question." "But the previous question was about pencils. Maybe it's different with buttons?" They laugh. But not because the question is pointless. On the contrary. It contains the secret of mathematics — abstraction. It does not matter if the objects in question are pencils, buttons or apples. The answer is the same in all cases. This is why we can abstractly say: $3 + 2 = 5$.

This is an elementary, but characteristic, example: Mathematics abstracts thinking processes. Obviously, every thought is abstract to a certain degree. But mathematics is unique in that it abstracts the most elementary processes of thought. In the example of $3 + 2 = 5$, the process involved is the joining of objects: 3 objects and 2 objects. One can ask many questions about these objects: Are they pencils or apples? Are they in your hand or on a table? And if they are on a table, how are they arranged? Mathematics ignores all these details, and asks a question that relates not to the various details, but only to the fact that these are objects that are joined: the resulting amount. That is, how many objects are there?

Abstract thought is the secret of man's domination of his environment. The power of abstractions lies in the fact that they enable us to cope efficiently with the world. In other words, they save effort. They enable going beyond the boundaries of the "here and now" — something discovered here and now can be used in another place and at another time. If 3 pencils and 2 pencils equal 5 pencils, the same will be true for apples, and it will also be true tomorrow. A one-time effort provides information about an entire world.

If abstractions in general are useful, then all the more so is mathematics, which takes abstractions to their limit. Therefore, it is not surprising that mathematics is so useful and practical.

Should Everyone Learn Mathematics?

People, on learning that I am a mathematician, often react with a thin smile, barely hiding a grimace of agony: "Mathematics wasn't one of my strong subjects." For so many people learning mathematics is such a tormenting experience that each generation asks the same question — what for? Why is this torture necessary? Shouldn't most people just give up on the attempt to learn mathematics? Nowadays, when a calculator can instantly perform mathematical operations, what is the point of learning the multiplication table or long division?

One answer is that mathematics is the key to all professions demanding knowledge of the exact sciences, and there are many of those. But mathematics is important not only for understanding reality. It offers much more than that — it teaches abstract thought, in an accurate and orderly way. It promotes basic habits of thought, such as the ability to distinguish between the essential and the inessential, and the ability to reach logical conclusions. These are some of the most significant assets that schooling can provide.

The question still remains unanswered — why is it so difficult? Must mathematics be a cause of suffering? A currently popular answer is "no" — the problem lies in the teaching. Common opinion is that many children considered to be "learning disabled" are actually "teaching disabled." But it can't be that simple. Blaming the teachers is too simplistic, and unreasonable. Anyone who claims that for hundreds and thousands of years mathematics teachers have been doing a bad job, must explain why this is so and why it isn't so in other subjects.

The special problem in teaching mathematics lies in the difficulty of conveying abstractions. You can tell people the name of the capital of Chile, but you can't abstract for them. This is a process each person must accomplish on his or her own. One must mentally pass through all the stages from the concrete to the abstract. The teacher's role in this process is to guide the student so that he experiments with the principles in the correct order. This is not a simple art that is easy to come by. But neither is it impossible. One of the purposes of this book is to relay some of the principles along the path of such "midwifery" teaching, as Socrates put it.

The Three Mathematical Ways of Economy

I didn't have time to write you a short letter, so I wrote a long one.

Blaise Pascal, Mathematician

Mathematics is being lazy. Mathematics is letting the principles do the work for you so that you do not have to do the work for yourself.

George Pólya, Mathematician

The true virtue of mathematics (and not many know this) is that it saves effort. This is true of any abstraction, but mathematics has turned economy of thought into an art form. It has three ways to economize: order, generalization and concise representation.

Order

Carl Friedrich Gauss was the greatest mathematician of the 19th century. One of the most famous stories in the history of mathematics tells of how his talent came to light when he was seven years old. One day, his teacher, looking for a break, gave the class the task of summing up all the numbers between 1 and 100. To his surprise, young Carl Friedrich returned after a few minutes, or perhaps even seconds, with the answer: 5050.

How did the seven-year-old accomplish this? He looked at the sum he was supposed to calculate, $1 + 2 + 3 + \cdots + 98 + 99 + 100$, and added the first and last terms: 1 and 100. The result was 101. Then, he added the second number to the one before last, that is, 2 and 99, and again the result was 101. Then 3 and 98, which yielded 101 again. He arranged all 100 terms in 50 pairs, the sum of each equaling 101. Their sum was thus 50 times 101, or 5050.

What little Gauss discovered here was order. He found a pattern in what seemed to be a disorganized sum of numbers, and the entire situation changed — suddenly matters became simple.

Imagine a phone-book arranged by a random order, or an unknown order. To find a phone number, you would have to go through each and every name. The order introduced into the phone-book, and the fact that we are familiar with it, saves a great deal of effort. A relatively small effort invested in alphabetical arrangement is returned many times over.

Or, think how much easier it is to live in a familiar city than in a strange one. A local knows where to find the supermarket or the laundromat. Knowing the order of the world around us provides us with orientation. Science, and mathematics in particular, has taken upon itself to discover the order of the universe, so that we may adjust our actions to it.

Generalization

There are many jokes about the nature of mathematics and mathematicians. The following is probably the best known of them all. I make a point of telling it to my students in every course I teach, since it is not only the most familiar, but also the most useful. It illustrates the principle of mathematical practice: Something once done does not require redoing.

How can you tell the difference between a mathematician and a physicist? You ask: Suppose you have a kettle in the living room. How do you boil water? The physicist answers: I take the kettle to the kitchen, fill it with water from the tap, place it on the stove and light the fire. The mathematician gives the same answer. Then you ask: Suppose you have a kettle in the kitchen. How do you boil water now? The physicist says: I fill the kettle with water from the tap, place it on the stove and light the fire. The mathematician answers: I take the kettle to the living room, and this problem has already been solved!

This brings economizing of thought *ad absurdum*, by placing it before true economization.

"This has already been solved" was also the answer we heard from the children who said they did not need to check how many buttons are 3 buttons and 2 buttons, since they had done the same with pencils. It appears, whether overtly or hidden, in each mathematical proof, and in every mathematical argument. "We have already done this, and now we can use it." In fact, this idea lies behind every abstraction: What we discover now will also be valid in other situations.

Proof in Stages: Induction

There is a mathematical process that is based entirely on the principle of "this has already been done." It is called "Mathematical Induction." A certain point is established in stages, with each stage relying on its predecessor, that is, on the fact that the previous case "has already been solved."

We will encounter this process several times throughout the book but will not mention it explicitly. For example, the decimal system is inductive: First, ten single units are collected to equal a new entity called a "ten." Then ten tens are collected to equal a new entity called a "hundred," and so forth. Yet another example is calculations: All algorithms used to calculate arithmetical operations are based on induction.

Concise Representation

The third mathematical economy is in representation. We are so used to the way numbers and mathematical propositions are represented, we forget that methods of representation were not always so sophisticated, and it was not so long ago, relatively speaking, that mathematical notation was much more cumbersome.

Let's begin with the representation of numbers. Up until about three thousand years ago, numbers were represented directly — "4" was represented by four markings, for example, four lines. This is a good idea for small numbers, but impractical for larger ones. Using the decimal system, we can now represent huge numbers concisely: A "million" only requires seven digits.

The second type of economy is in the representation of propositions. A "mathematical proposition" is the equivalent of a sentence in spoken language. Up until a little over two thousand years ago, mathematical propositions were phrased in words, for instance: "Three and two is five." Then, a very useful tool was invented: the formula. Its originator was probably Diophantus of Alexandria, who lived during the 3rd century B.C. Formulas are not only shorter, they are also more accurate and uniform, and allow systematic handling.

Historical Note

The notation we currently use developed slowly and gradually. Its current form was only established during the 16th and 17th centuries. The sign of equality ($=$), for example, only appeared in the mid-16th century. Its inventor, the Englishman Robert Recorde, explained his choice by saying that "no two things can be more equal than a pair of twin lines of one length."

Summary: Mathematical Economy

Mathematics has three ways of saving effort:

- *Order*: Discovering a pattern makes orientation easier.
- *Generalization*: A principle discovered in one area can be applied to other areas.
- *Concise Representation*: The decimal system is a wonderfully economic way of representing numbers; mathematical formulas represent propositions in a brief and clear manner.

The Secret of Mathematical Beauty

Euclid alone has looked on beauty bare.

Edna St. Vincent Millay, Poet

If the solution is not beautiful, I know it is wrong.

Buckminster Fuller, Architect and Inventor

In one second-grade class, I showed the children an elegant way of proving the commutative rule of multiplication (we will encounter it in the chapter about the meaning of multiplication). A child sitting in the first row looked ahead for a moment, and then said quietly: "It's beautiful."

Ask a mathematician what it is about his profession that appeals to him, and nine out of ten times the answer will be "beauty." Mathematics is useful in everyday life, but to those who deal with it that is not the essence. To them its main characteristic is its beauty. A mathematical discovery rewards its owner, and those who study it, mainly with aesthetic satisfaction. What does mathematics have to do with beauty? What possible relation could there be between the cold and dry subject of mathematics and the beauty found in art?

This brings us to the notoriously difficult and intricate question, "What is beauty?" It may actually be mathematics, the uninvited guest in this arena, that can shed light on the answer. For it is there that we find a pretty unanimous agreement on the meaning of beauty: A mathematical idea is beautiful when it introduces a new and unexpected element, one that appears as if from nowhere. The person who invented the decimal organization of numbers undoubtedly felt a powerful sense of beauty. The first person to discover the possibility of summing numbers by writing them one above the other surely had a sense of aesthetic satisfaction.

The Great Book in Heaven

The famous Hungarian mathematician, Paul Erdős, used to talk of the "Great Book in Heaven" containing the most elegant proof for every theorem. I believe many would agree that the first page of the Book should be dedicated to Euclid's proof that there are infinitely many prime numbers. Not only is it one of the most elegant proofs known in mathematics, it is also one of the most ancient.

14

A prime number is a number that is not divisible by any other integer (that is, integral number), except for 1. For example, 2 is a prime number, as are 3 and 5. The integer 4 is not prime, as it is divisible by 2. Any integer can be written as a product of prime numbers (not necessarily different ones: 4, for instance, is 2 times 2). The first five prime numbers are 2, 3, 5, 7 and 11.

Of course, 11 isn't the last prime number. For example, 13 is a larger prime number. But Euclid claimed that he could prove there was a prime number greater than 11, without knowing what it was! The idea is simple. Look at the product of the first five prime numbers, namely, $2 \times 3 \times 5 \times 7 \times 11$. The result, 2310, is obviously divisible by 2, 3, 5, 7 and 11. Therefore, the integer 2311, the result of adding 1 to 2310, is not divisible by any one of these numbers (if an integer is divisible by 2, its successor is not divisible by 2; if it is divisible by 3, its successor is not divisible by 3, etc.). But, like any other integer, 2311 is divisible by some prime number, and, as mentioned before, it cannot be 2, 3, 5, 7 or 11 since 2311 is not divisible by these. Therefore, there must be a prime number that is not one of these numbers.

The same will obviously apply if we take the first 100 prime numbers. What we demonstrated is that for each prime number there is a greater prime number. Therefore, there are infinitely many prime numbers!

This proof is doubly elegant. First, because of the idea, appearing as if "from nowhere," of multiplying all prime numbers up to a given integer and adding 1. Second, because the indirect proof demonstrates the existence of a greater prime number without actually naming it.

"Knowing Without Knowing"

We still haven't arrived at the root of the matter — what inspires a sense of beauty? In art, for example, beauty isn't necessarily derived from the introduction of unexpected ideas! Is there a relationship between mathematical beauty and the beauty of poetry, for example, or that of music?

To understand this relationship, consider the power of poetic metaphors. The beauty of a metaphor is derived from its indirect message, something said without actually being said, so that the receiver does not have to look it straight in the eye. Take for example the following metaphor from the Song of Songs:

> *Look not upon me, that I am swarthy, that the sun hath tanned me; my mother's sons were incensed against me, they made me keeper of the vineyards; but mine own vineyard have I not kept.*

> (Chapter 1, Verse 6).

The metaphor in the last line contains a simple message, but for a moment the reader can pretend not to understand it, as if it is truly about a vineyard that is not well-guarded.

What happened here? As in mathematics, an "idea from somewhere else" suddenly appeared — a vineyard instead of an erotic message. As a result, we perceive the message on one level, while not completely absorbing it on another level. This is "to know without knowing." Likewise, in a surprising mathematical solution the unanticipated connection of ideas enables us to understand the newly discovered order on one level, while the ordinary tools of reason, still using the old concepts, do not suffice to grasp it.

Is this true of all types of beauty? I believe so. Rare beauty is wondrous in our eyes. It contains something that we do not fully understand. For example, a magnificent view inspires a sense of beauty because it lies beyond the scope of our ordinary tools of perception.

Mathematics and Art

> *It is true that a mathematician who is not also something of a poet will never be a perfect mathematician.*

> Carl Weierstrass, Mathematician

Mathematics has two characteristics in common with art: order, and economy and concision. Art, like mathematics, finds order in the world. Music, for instance, is organized noise, while paintings create order in the visual experience. As for concision, poetry, for example, is famous for shortening and compressing many ideas into one saying. In German, the word for poetry is "Dichtung," meaning "compression." The poet Ezra Pound defines great literature as *language charged with meaning to the utmost possible degree.*

These are all related to "knowing without knowing." Order inspires a sense of beauty when it is perceived by the subconscious. There are two possible explanations for such a mode of perception: Either the order is so surprising that standard perception does not keep up with it, or it is too complex to be perceived by reason. Economy and concision also have the effect of "knowing without knowing" — the idea flies past us so quickly that we do not have time to grasp it. The same is true of the compression of several meanings into one expression — it does not enable conscious comprehension of all the significations.

The elementary arithmetic we learned as children contains some of the most beautiful mathematical discoveries ever made. Why, then, is it not perceived by most people as beautiful? Mainly because it is often learned mechanically, in a way that does not reveal its beauty. But it is not too late, and those who are able to see the principles of elementary mathematics in a new light will be able to rediscover their beauty. I can testify that this is what happened to me.

Layer upon Layer

Professors demonstrated free thought/And thoughts of gymnastic instru-
ments, in groups and individually/But most of their words remained unclear
to me/I probably was not yet ready.

Yehuda Amichai, "I Am Not Ready", **Poems**

Fermat's Narrow Margins

In 1637, the French mathematician, Pierre Fermat, wrote a conjecture in the margins of a book, a copy of Diophantus' *Arithmetica*: "I have discovered a truly marvelous proof," he added, "but the margin is too narrow to contain it."

Generations of mathematicians were tormented by the thought that the proof of what became the most famous mathematical conjecture of all times was truly lost, and dedicated themselves to reproducing it. After a while, it became clear that Fermat, like many others that followed him, was deluding himself, and was actually mistaken in his proof. When a proof was finally found in 1995 by the Englishman Wiles, it became obvious that it could not have fit into the margins of a book. It was 130 pages long, and if added to the many arguments on which it was based, would fill thousands of pages.

Shorter proofs than the one to Fermat's conjecture are also constructed of many layers, each one based upon the other. Each layer is established in turn and serves as a basis for the next, according to the "this has already been done" principle. There are other fields in which knowledge is built on previous knowledge, but in no other field do the towers reach such heights, nor do the topmost layers rely so clearly on the bottom ones.

The first fact one must know about mathematical education is that this is true not only of advanced mathematics, but also of elementary mathematics. There, too, knowledge is established in layers, each relying on the preceding one. The secret to proper teaching lies in recognizing these layers and establishing them systematically.

A famous anecdote in the history of mathematics tells of the impossibility of shortcuts. The hero of the story is Euclid, who lived in Alexandria between 350 and 275 B.C. and authored *The Elements*, the most influential geometry book of ancient times (and possibly of all time). Among other achievements, he defined in it the terms "axiom" and "proof," two of the greatest achievements of mathematics. Ptolemy, the king of Egypt at that time, asked for his advice on an easy way to read the book. "There is no royal road to mathematics," replied Euclid. Even kings cannot skip stages.

Note: The 5th century Greek historian, Stobaeus, attributes the same story to different characters: Alexander the Great and his teacher, Menaechmus.

The same is true of elementary mathematics. However, since it deals with the bottom of the tower, the number of layers it establishes is smaller. There are no long chains of arguments as in higher mathematics. This is one of the reasons it is appropriate for children. In another sense, though, it is harder. Some of its layers are hidden and difficult to discern, as if they were built underwater and thus difficult to view. Noticing them requires perceptive observation. They are easy to miss and skip. Elementary school mathematics is not sophisticated, but it contains wisdom. It is not complex, but it is profound.

Mathematics Anxiety

Education researchers use the term "mathematics anxiety." There is no history anxiety, or geography anxiety, but there is mathematics anxiety. Why?

The main reason lies in its layered structure: Mathematics anxiety arises when one stage is unheedingly skipped. As mentioned before, many of the layers of mathematical knowledge are so elementary that they are often easy to miss. When this happens and an attempt is made to establish a new layer on top of the missing one, neither the teacher nor the student can discern the origin of the problem. The student hears something that is meaningless to him, since he is "probably not yet ready." The teacher is also perplexed, since she cannot identify the source of the difficulty. When one does not understand the origin of a problem, unfocused fear arises and anxiety is born.

A "layer" need not be an explicitly stated piece of information. Sometimes it is the acquisition of experience. For example, to acquire the concept of the number, one must have extensive experience in counting. Something happens in the mind of a child when he is counting.

Something is gradually built, requiring investment of time and effort even if the results are not immediately apparent.

One cannot mention mathematics anxiety without also mentioning the other side of the coin — the joy of mathematics. Just as anxiety is not associated with any other subject, so too the happiness that beams from a child's face when he understands a mathematical principle is not seen in any other subject. As likely as not, there is a connection between the two.

An Example of the Importance of Not Skipping Stages

Here is an example from my personal experience of what happens when
you skip a stage. An experienced teacher would probably not have fallen
into the trap as I did. She would have known how difficult the term I was
trying to teach, that of "greater by... than... " or "more by... than... ",
is for children. But that trap turned out to be an instructive lesson for me.
I learned the importance of establishing concepts in the proper order, and
how far one can go when this is done.

For a certain period of time, I taught two first-grade classes in Maalot.
One day, nearing mid-year, I arrived with the intention of teaching both
classes problems which included expressions such as "greater by 4 than..."
or "4 more than..." In the first class, I wrote on the blackboard: "Donna
has 4 pencils more than Joseph. How many pencils does Donna have if
Joseph has 5 pencils?" The order in which the data were presented was not
incidental. I deliberately began with the relation between the number of
pencils Donna and Joseph had, and not with the absolute number of pencils
Joseph had. I wanted them to understand that it is possible to discuss the
relation between the numbers without knowing the point of reference (how
many pencils Joseph has).

Up to that point, the children had no difficulty translating real-life
stories into arithmetical expressions. This time was an exception. Confusion
prevailed in the classroom. I tried to phrase the question directly: "Joseph
has 5 pencils, Donna has 4 more pencils. How many pencils does Donna
have?" But this didn't help either. Most of the children were not following.

By that time I understood that I had skipped a stage. As a matter of
fact, I had skipped more than one. It was not only the concept of a yet
unknown relation between two elements that was difficult for the children.
It was also the concept of having 4 more than or greater by 4 than itself
that was unfamiliar, and a first-grade teacher should have been well aware
of that fact. This is not a concept the child encounters in his everyday life.
Most children are familiar with the term "greater than" but not necessarily
with the meaning of "greater by 4."

There was no other way but to start anew. I drew a large circle and a
square on the blackboard, and asked the children to draw an equal number
of triangles inside the square and the circle. Then I asked them to add one
triangle to the circle, and asked which shape contained more triangles. Then
I asked how many more. At this point the lesson ended.

From this lesson I moved on to the other first-grade class. By now I was wiser and began the lesson properly, from the beginning. I invited two children to the blackboard and gave each 5 crayons. I asked which child had more, and was told that they both had the same number. I gave one child an extra crayon and asked: Who has more now? How many more? By how many does the other child have less? I gave the first child yet another crayon and repeated the same questions. I continued to give the child more crayons, and asked at each stage how many more crayons he had than the other child, and by how many the other child had less. Then I gave the second child more crayons, one by one, until they both had an equal number. The next stage was going in the opposite direction — taking away one crayon after another from one child, and asking at each stage who had more, and by how many more. Throughout the entire process, without giving in, we also asked who had less, and by how many less.

I then drew a set of stairs on the blackboard and numbered them. I drew two children: one on Stair 9, the other on Stair 6. I asked the class how many stairs the child positioned lower needed to climb to reach the child on the higher stair. Then, how many stairs the higher child needed to descend to reach the lower one. I asked: "By how many stairs is the first child higher up than the second?" and "By how many stairs is the second child lower than the first?" We went through several similar examples.

Now we took the abstraction one step further. Instead of the concrete stairs, we switched to age differences. I asked one child by how many years he was older than his brother. By 3 years, he answered. By how many years was his brother younger than him? From then on, we had a ball! I asked: How old would he be when his brother was 20 years old? And how old his brother would be when he was 100 years old? And when he was a 1000 years old? Some of the children followed me through to the high numbers — in almost every first-grade class there are students who can calculate in the hundreds, even complicated calculations like $1000 - 3$. Of course, they were very amused by the thought of what would happen to them in a hundred or a thousand years.

The final part of the lesson was dedicated to personal experimentation. We used improvised abacuses, wooden skewers on which we threaded beads made of play-dough. (These are better than regular beads since the children make them themselves and can feel the material in their hands. The play-dough beads also don't roll noisily on the floor.) The children paired up, and I asked that in each pair one child would put 3 more beads on the

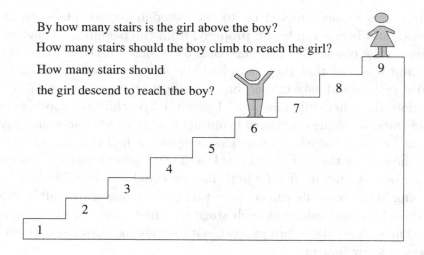

By how many stairs is the girl above the boy?

How many stairs should the boy climb to reach the girl?

How many stairs should

the girl descend to reach the boy?

skewer than the other child. This was another source of enjoyment. I didn't tell them how many beads each child should thread, or who should have more beads. I only mentioned the difference, and this drove them to higher numbers. In one pair, for example, one child threaded 10 beads, confident that he would be the winner. When he found out he was beaten since his partner had threaded 13 beads on her abacus, he added 6 beads to his. A well-known pedagogical dictum is that a lesson should go through three stages: the concrete, to drawing, and finally to abstraction. In this sense, the lesson was exemplary: We began with the concrete (the crayons), moved on to the drawing of the stairs, and finally on to the discussion of the ages of siblings, in which the numbers were not concretely presented. Finally, we ended the lesson with an active implementation of the concepts we had learned (play-dough beads and wooden skewers).

So much for didactics. What about content? What conceptual structures did the children acquire during this lesson? The answer: more than first meets the eye. First of all, they learned the concept of relation, like "greater" or "smaller," between numbers. Furthermore, they understood the message I was trying (unsuccessfully) to convey in the first class: discussing a relation between numbers without having the actual numbers at hand. The assignment they received at the end of the lesson was to create a situation where one child had 3 beads more than his or her partner, without being told explicitly how many beads each child should have.

In addition, the children learned that a relation can be viewed from both directions and that matters look different from each angle: If one is greater by 3, than the other is smaller by 3. They realized that the relationship changes when you change one of its components. (If I have 3 more, you can have the same if you add 3 to your own, or if I subtract 3 from mine.)

Of course, they also learned the concept of "greater by..." itself, which was the purpose of the lesson. And right along with it, the relationship between "greater by..." and addition: If you add 4 to a number, the result is greater by 4 than the original number.

Another idea conveyed in this lesson was the law of conservation. "Conservation" means that something remains constant while other things change. For example, when you rotate a triangle, its position changes, but the relationship between its sides remains constant; the angles are

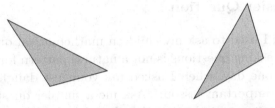

A conservation law: The angles are preserved when the triangle is rotated

the same. This lesson taught conservation of difference: The difference between the ages of two brothers is always the same. If you are 7 years old and your brother is 4 years old, the difference between your ages is 3. 23 from now, the difference will remain unchanged: 27 is greater than 24 by 3. That is, if you increase the numbers by the same amount, the difference between them will remain constant. This law will accompany the children throughout their school years. They will encounter it, for instance, when expanding fractions, where the ratio between two numbers does not change if you multiply them by the same number.

Law of conservation: 17−14=7−4

* * *

The first of the two lessons always serves me as a reminder of: how elusive the structures of thought that are established at an early age really are; how easy it is to believe that they exist in children's minds, and how easy to forget that even the most elementary structures need to be established at some point, through hard work.

But it was the second lesson that was truly instructive, and there I learned a lesson which will reappear throughout this book. I realized that by establishing concepts in the proper order, and teaching them through concrete experimentation, you can go a long way. The concept of "greater by ..." was established through concrete experimentation, dwelling upon each of its aspects, even those that seemed most simple. We insisted on explicit phrasing, even for the most obvious principles. Through these we managed to go far beyond what I had dared to even dream of at the beginning of the lesson.

Ask Me an Easier Question

As a fresh parent, I used to ask my children mathematical questions. I do it less these days — asking questions is not a natural pattern for a parent–child relationship. But one day, when I asked my youngest daughter a question, she taught me an important lesson: "Ask me a simpler question," she said. She was not trying to evade the question, but asked me to provide her with a preliminary step. If a question is hard, it usually means that some previous stage is missing.

I use this phrase in my classes. When students find a problem too hard, I tell them the story of my daughter, and encourage them to do the same: Whenever they fail in solving a hard problem, they should ask for a simpler one. My aim is to make them aware of the possibility that there is a stage missing in their knowledge. This not only averts frustration, but also makes them aware of their thinking processes, an important goal by itself.

Step by Step

One day I was watching a small class of weaker students. Some of them had a hard time calculating sums like 8+6. They were studying conversion between hours and minutes. The teacher asked them the following question: There were three meetings of 50 minutes each. How much time did the meetings take altogether, in terms of hours and minutes? I knew well what would have happened if free discussion were to follow: One or two students would know the answer, the rest would be left behind. So I asked the teacher's permission to step in. I told the students that in mathematics one has to go very slowly, and then I told them the story of my daughter (who was about their age) and her "give me a simpler question" request. I said that I was going to ask them questions in steps, and promised that each question will be simple. Then I asked a girl who beforehand refused to answer questions how many hours are in 60 minutes. This she knew — one hour. Then I asked how many hours and minutes there are in 61 minutes. That was not hard, too. Then I asked about 62 minutes, and 63, not skipping a stage. This was slow, but all the kids were with me and all had a sense of achievement. When we got to 90 minutes, I dared to make a jump. And now, I asked, how about 100 minutes? And 110? Then we slowed things down again: 111, 112, until we got to 119 minutes — an hour and 59 minutes. Then we got to 120 minutes, which they said, as I had expected, is one hour and 60 minutes. But what are 60 minutes, I asked, and they got to the desired answer: 120 minutes is two hours. From there on to the number appearing in the original question, 150 minutes, it was easy and all the students knew the answer.

Everybody can make forward steps if they are small enough. One only has to know how to break the problem into small steps, and how not to be in a rush. In the long run, going in small steps saves time, not wastes it.

Divide and Conquer

An error no less common than skipping stages is teaching two ideas (or more) at once. Ideas should be taught separately, even if they are not dependent on each other, and even if the order in which they are taught is immaterial. It is important to teach each stage individually. Divide and conquer, or in other words, "break the principles into components," is one of the principal rules of good teaching.

In many cases, breaking a problem into stages is all that is required — the child will do the rest himself. For example, when a child finds it difficult to calculate 2×70, it is sometimes enough to ask him how much is 2×7 and he will complete the missing information on his own. It is enough to provide a person with an intermediary step on the ladder, and he will climb it himself. This reminds me of a saying I once heard from a teacher of mine: "A mathematical proof is a non-trivial combination of trivial arguments." The difficult part is breaking a problem into interlinked stages.

Whole Numbers

God made the whole numbers; all else is the work of man.

Leopold Kronecker, Mathematician

Why were Numbers Invented?

My 9-year-old daughter likes to sign birthday cards with the words: "With lots and lots and lots... of love," filling half the page with "lots."

It could be that thousands of years ago, before numbers were invented, this method was used as a substitute for counting. Instead of saying "3 stones" the caveman would say: "stone, stone, stone."

Now it is clear why numbers were invented: to economize! Instead of filling up half a page, my daughter could write "a hundred times lots of love."

Though I must admit it doesn't have the same effect.

* * *

In one of my third-grade classes, I wanted the children to understand how economic the use of numbers is. So, I told them a story. "The story I am about to tell you," I warned them in advance, "happened before something was invented. Can you guess what that something is?"

A caveman returned to his cave after a day of hunting, and said to his wife: "I brought you a rabbit, a rabbit, a rabbit and a rabbit." His wife answered: "Thank you, thank you and thank you."

The children had no trouble guessing: The story took place before the invention of numbers. Nowadays we would say in short "4 rabbits," or "3 thank you's" (or, as is more customary, many thanks).

There is no problem saying "rabbit, rabbit, rabbit and rabbit" when there are 4 rabbits. But just imagine what would have happened if the caveman had brought 100 rabbits! Numbers save a lot of work, utilizing all three ways of mathematical economy: representation, generalization (the number 4 is used to count rabbits, pencils and cars) and order — knowing how many items there are of each kind provides essential information about the world and creates a certain order.

The numbers, 1, 2, 3, ... , were born of the fact that the same type of unit can be repeated several times. Since their invention was so natural, they were given the name "natural numbers." All other types of numbers were invented at a later stage, and they are indeed less natural since they are more distanced from real life: fractions, negative numbers, real numbers, complex numbers, and so on.

In this book, when the word "number" appears on its own, it refers to natural, namely whole numbers.

Why Do Numbers Play Such a Central Part in Mathematics?

Ask a passerby what is mathematics, and the answer will probably include the word "numbers:" Mathematics deals with numbers. Professional mathematicians know this isn't accurate. Certain mathematical fields, such as geometry, do not deal directly with numbers. Still, there is much truth to the popular view: Numbers do indeed have a special role in mathematics. They appear in almost every mathematical field, at least indirectly. Why is this so?

Mathematics, as mentioned before, abstracts the elementary processes of thought. Numbers play such a central part in it because they are a result of the abstraction of the most elementary of all processes: sorting the world into objects. Man recognizes a part of the world, separates it from the rest, defines it as a single unit, and gives it a name: "apple," "chair," "family." This is how words were created, and how the number "1" was born — "1 apple," "1 chair," "1 family." Natural numbers came about through repetition of a unit of the same kind: "2 apples, 3 apples ... "

Numbers with a Denomination and Pure Numbers

"I planned on having one husband and seven children, but it turned out the other way around."

Lana Turner, Movie Star

Numbers are important, explains Lana Turner. But first and foremost is the denomination, namely, what the number counts.

The concept of the number did indeed begin with numbers that have denominations, that is, with the counting of objects. The abstract number came along at a later date, born of the fact that important properties of numbers, such as the results of arithmetical operations, are not dependent on the denomination: $2 + 3 = 5$ is true of apples and chairs alike. Therefore, man abstracted: From 4 apples and 4 chairs he invented the "pure" number, that is, a number without a denomination: 4. Pure numbers can be used to phrase propositions that are true of any object.

It is the concrete examples that lead to abstraction, and therefore numbers with denominations should be taught before pure numbers. In other words, the number should be taught through counting existing objects. A first-grader should count as much as possible. This is the only way to establish the concept of the number. A first-grade classroom should be full to the brim with buttons, beads, popsicle sticks, straws. When counting, the denomination should always be mentioned. The answer to the question: "How many pencils do we have here?" is not merely "4," but "4 pencils."

First Lesson on Denominations

The following is a suggestion for a first lesson on denominations. Warn the class that you are about to give them strange instructions. When you have their attention, ask one of the children: "Give me two." The ensuing discussion will teach the children that when you say "two" you must also say two of what.

Sets

The first arithmetical operation is neither addition nor subtraction, but that which gave birth to the number — the definition of a unit. In other words, it is the separation of an object from the rest of the world, giving it a name and identifying it as "a unit," which can then be repeated several times.

In fact, the definition of a unit is more basic an operation than the invention of numbers. It serves to establish other mathematical concepts as well. In particular, it is the basis for a most important mathematical concept, second only to the number — the concept of the set. This concept originated in the finding that a few items can be grouped to form a new unit, called a "set." Several people group together to form a set called a "family" and 5 players form a basketball set (or team).

This operation is most prominent in the decimal system used to organize numbers. Ten terms are grouped to create a new unit called a "ten." Ten tens can then be used to create another unit, a "hundred."

Order

The numbers are used not only for counting, but also for ordering objects: "first, second, third,..." Numbers can be compared and ordered according to this comparison. When a number a is smaller than a number b, we write: $a < b$. For example, $3 < 5$. Some children find it difficult to remember which direction the sign goes, and most textbooks provide a mnemonic: The larger, open side of the sign "$<$" faces the larger number. The sign "$<$" is used only for pure numbers; we do not use the notation "3 apples < 5 apples." Therefore, in first grade, when the numbers are mainly used with denomination, it is better to use the terms "more" and "less," or the more officious "greater than" and "smaller than," and leave the "$<$" notation to the second grade.

There is a basic tool for visualizing order among numbers. It is called the "number line." It is a straight line, with the numbers marked on it at evenly spaced intervals. Again, this is used for the order among pure numbers, and therefore its study is better deferred to the second grade. First-graders are still in the counting stage, and the even spacing, as well as the association with the straight line, and the identification of the numbers with points, is too abstract for them.

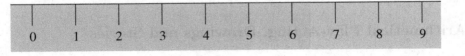

Meaning and Calculation

Arithmetic in Search for Meaning

Most people associate "arithmetic" with arithmetical operations, and arithmetical operations with their calculation.

The first association is definitely correct. Arithmetic does indeed deal mainly with arithmetical operations. In contrast, the second association is far from being true. The operations and their calculation are not one and the same. Calculation is only the second stage. First and foremost is understanding the meaning of the operations.

The meaning of an operation lies in its link to reality: which situations require addition of numbers, which require subtraction and which multiplication or division. For example, the meaning of addition is to join: $3 + 4$ is achieved by joining together 4 objects and 3 objects. The meaning of subtraction (amongst others) is to remove: $7 - 3$ corresponds to a situation where 3 out of 7 objects are removed.

It may sound simple, but in fact therein lies most of the depth of arithmetic, mainly because the rules that guide the use of operations are derived from their meaning.

Arithmetical Play-Acting, Drawings and Stories

As with any abstract concept, the road to the meaning of operations begins with the concrete. To understand addition, the child must experience joining sets of objects over and over. To understand division, he must experiment in dividing a set of objects into equal sets.

There are three ways to experience the arithmetical operations: play-acting, drawings and stories. An especially efficient way of initially introducing an arithmetical operation is arithmetical play-acting. At first, the teacher is the director. Two children are invited to the front of the classroom; one is given 3 straws and the other 2 straws. How many do they have altogether? After the children reply "5 straws," point to the set of 3 straws, the set of 2 straws, and say out loud: "3 plus 2 equals 5. Let's write it down." But remember — arithmetic likes to be concise. Instead of words, we use signs — and now write the plus sign on the blackboard: $2 + 3 = 5$. Then, a different variation: One child holds 4 pencils in his hand and the other holds 3 pencils. The first child gives all his pencils to the second child. How many pencils does the second child now have? What is the appropriate mathematical expression? (The term in general use is "sentence," or "mathematical sentence.") How many pencils does the first child now have? The same play can be repeated, only this time the second child gives all his pencils to the first.

In the next stage, the children become the directors. They decide on the arithmetical play, play in it and write the mathematical sentences on the blackboard.

Then come the arithmetical drawings. The teacher can demonstrate by drawing 3 flowers alongside 2 flowers. How many are there altogether? Then the children should draw on their own. First on the blackboard and then in their notebooks or personal boards. For addition, the drawings are simple; they become more complex with subtraction and then multiplication and division. Here's a tip for subtraction drawings: When you demonstrate, for instance, $5-2$ by drawing 5 balloons and removing 2 of them, don't erase the 2 you remove, but just cross them out. This way, the subtracted amount can still be seen.

The last stage is that of telling arithmetical stories using words. "Zack has 3 flowers, Amber has 2 flowers. How many do they have altogether?" This is already abstract, because the numbers are not represented by objects or drawings, but by their names.

Inventing Your Own Arithmetical Stories

I hear — I forget. I see — I remember. I do — I understand.

Confucius

You can't learn how to drive by watching others drive, and you can't learn to dance by watching *Swan Lake*. Similarly, the meaning of the operations cannot be fully understood just by hearing arithmetical stories from others. You must be able to invent such stories on your own.

The ability to do so is the true test of understanding the meaning. Tell a story that matches the expression $3+4$, or $4-3$ ("Dina received 4 arithmetic exercises as homework. She solved 3. How many does she have left?"), or 3×4, or $12\div3$.

Inventing arithmetical stories has an additional advantage: It teaches the reversibility of processes. First, we made the transition from an arithmetical story (Amber has 4 arithmetic exercises, she solved 3, how many does she have left?) to an arithmetic exercise $(4-3)$. Now we learn that the opposite is also possible: Given the exercise $4-3$, a matching story can be invented. In fact, more than one story.

Calculation Means Finding the Decimal Representation of the Result

What is "calculation?" Figuring out the result of an exercise, of course. But this is only a partial answer, which does not touch on the main point. For the last millennium most of mankind has been representing numbers using the decimal system, and hence the essence of calculations is in figuring out the decimal representation of the result.

Before the decimal system was invented, calculators had a simple life. The number 4, for example, was represented by ||||. The meaning of a calculation such as 8 + 4 was to draw 8 sticks, and 4 more alongside them. The result was also written using the same marking, like this:

$$|||||||| + |||| = ||||||||||||.$$

Is there any sophistication to such calculations? None at all. The caveman did not need to send his children to school to learn this. All the knowledge required here is that the meaning of addition is "to join" — the result is achieved by joining the two sets of lines. No calculation is required.

Nowadays the same exercise is written differently: 8 + 4 = 12. Is there any cleverness in this? Is there any point in learning this in school? This time the answer is a decisive "yes." This requires a true operation: the grouping of a ten. Of the 12 lines in the result, ten are grouped into one ten. This calculation provides information: The result includes one ten and two ones. Something has been said here about the decimal representation of the result.

Calculation is figuring out the decimal representation of the result from the decimal representation of the problem's components. This is one of the reasons why knowing how to calculate is so important in elementary school. Its purpose isn't just to figure out the results of problems, but also to achieve a deeper understanding of the decimal system.

How Do We Calculate?

Most people remember from their arithmetical studies mainly the ways to calculate the operations. These are recipes for performing the steps, one by one, according to a predetermined order, much like a recipe for baking a cake. A fixed recipe, one that dictates actions in a certain order, is called an "algorithm."

The algorithms we were taught in school are so deeply ingrained in us that we tend to forget they are not exclusive. That is, there is more than one way to perform each arithmetical operation. The algorithms we are familiar with, those taught in school, make use of a pen and paper and are based on writing the number vertically, one above the other. Hence their names: vertical addition, subtraction and multiplication. Division received a different name — "long division." These algorithms were introduced in Europe by the Arabs during the 12th century, and were established slowly. The fact that these algorithms remained unchanged for hundreds of years to this day bears witness to their wisdom.

Historical Note

The word "algorithm" is derived from the name of a 9th century mathematician, Al-Khwarizmi, who was born in the region of Khwarizm (now part of Uzbekistan) and lived in Baghdad. The Europeans learned the decimal system, and methods of performing operations with it, from the translation of his book. Therefore, they called any calculation an "algorism," which out of confusion with the Greek word "arithmus" (number) later became "algorithm."

Our generation's calculation algorithms make use of a pen and paper. This was not always so. Between the 12th century, when the decimal system was introduced in Europe, and the 16th century, there was a competition between the "algorithmists," those who calculated using paper (as we do today) and the "abacusists," those who used an abacus. The decimal system became standard only after the algorithmists prevailed.

Why are Addition and Subtraction Exercises Written Vertically?

Ever since the first grade I have known that Hebrew is written on lined paper and arithmetic on graph paper. What I didn't understand then was why. This should be explained to the students — it isn't a state secret. The purpose of the squares is to make it easier to write the ones beneath the ones, the tens beneath the tens, and so forth. What for? So that we can add or subtract items of the same kind: ones with ones, tens with tens. It is simply a matter of a "common denomination." Take, for example, the exercise:

$$
\begin{array}{r}
23 \\
+\,64 \\
\hline
87
\end{array}
$$

The vertical writing makes it easier for us to understand that the 3 ones in 23 should be added to the 4 ones in 64, and the 2 tens in 23 to the 6 tens in 64.

The Decimal System

Ten fingers have I They can build anything.

Rivka Davidit, Ten Fingers Have I

Organization and Representation of Numbers

Anyone using large numbers needs a good system to organize and represent them. The system used today is the decimal system. This is undoubtedly one of the most ingenious and useful inventions ever made by man. It enables not only concise writing of numbers, but also simple and efficient calculation. It is our generation, the computer generation, that provides the decisive proof of its usefulness. Much effort is currently invested in researching ways to simplify calculations. If a more efficient method of representing numbers existed, it would probably have been discovered by now. The fact is that the principles of number representation in computers are still identical to the principles of the decimal system, with the only difference being that objects are paired instead of being grouped into tens.

Two Principles: Grouping Tens and Place Value

The decimal system is based on two principles. One relates to the organization of numbers, or more precisely, of object sets. The other relates to the writing of numbers.

The first principle is about collecting — grouping ten elements to form a new unit. This process is repeated: Ten ones are collected to form a ten, ten tens to form a hundred, ten hundreds a thousand, and so forth. The second principle relates to writing numbers, using a system called "the place value system." It means that the value of each digit is determined by its position in the number. The rightmost digit enumerates ones, the second digit from the right enumerates tens, the third from the right hundreds, and so forth.

> The decimal system is used to organize and represent numbers. It is based on two principles: grouping tens, and ascribing a digit with a value according to its place in the number. The more to the left the digit is placed, the greater its value.

> The decimal system simplifies writing and facilitates calculations of large numbers.

Numbers Were Not Born Organized

When I tried as an adult to learn to play ping-pong, the instructor discouraged me. He claimed I was too used to the wrong movements, and that it would be difficult to get rid of them. This problem doesn't usually occur in arithmetic, where prior knowledge is generally helpful. But there is one area in which the knowledge a child brings from kindergarten creates a problem: the decimal system. In first grade he must learn anew, in a different way. The reason is that the child already knows how to count, and to him the decimal system is as natural as if it were an integral part of the number. In fact, the same is true of adults. Most of us find it difficult to imagine representing numbers in any different way.

The most important thing to understand about the decimal system is possibly that numbers were not born naturally organized — we are the ones who organize them. 34 objects are not naturally clustered into 3 clusters of ten. We are the ones who group them into 3 tens and leave 4 single ones.

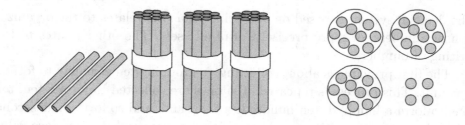

Why is this important to know? Because a child who understands that he himself puts together and takes apart the tens, at his own whim, no longer sees calculations using the decimal system as magic. He knows that when ten ones appear in a calculation, he can group them to form a ten. And more importantly, if he knows he is the one who grouped them, he also understands that he can take them apart when the need arises. The need to take apart tens does indeed arise, in subtraction.

This is also the reason why a child must experiment in grouping tens himself, without using instruments that perform the job for him. He must group objects into sets of 10. Have him thread 10 beads on a string, draw small circles on a page and ask him to group them into tens and draw a line around each group. If, for example, he drew lines around 4 tens, and 5 circles remained outside the lines, he should understand: There are 45 circles.

Second-graders should experiment in grouping hundreds. Ten bundles of ten matches each are grouped to form an element called "a hundred." Don't be afraid of using concrete examples of grouping even in the thousands. Once a child experiences this at least once, he develops a good sense of the fundamental nature of the decimal system.

Where Do We Begin?

When teaching, one should always begin with the familiar. In this case, the familiar is the child's knowledge of counting. He has a vague sense that when crossing the border at ten some sort of operation is performed — putting aside a ten and starting anew.

This knowledge now needs to be explicitly worded. The process of grouping tens, still vague, must be clarified. One way to do so is to count objects along with the children. After counting ten objects, tie them up together, or set them aside in a pile, and continue counting: eleven, twelve, thirteen . . .

Historical Note

Who Invented the Decimal System?

The system we are familiar with today was invented in India. But the priority actually belongs to the ancient Babylonians: They had already invented both principles (grouping and place value) some 3,700 years ago.

However, the Babylonians didn't group tens. They collected sets of 60. That is, 60 ones were grouped to form one element of 60, and 60 elements of 60 were grouped to form a new element, containing (in our notation, of course!) 3,600 ones. This method of writing was adopted by the Greeks, who used it for complicated calculations, such as those used in astronomy.

The use of the decimal system began, as mentioned, in India during the 6th or 7th century. The Arabs adopted the system in the 8th century, and introduced it in Europe during the 12th century.

The Arabs also adopted the way of writing digits from India. But the digits developed differently in different places. The Western world uses the

*digits that evolved in Arabic Spain. The Arabs themselves, on the other hand,
use the digits that evolved in the Middle East.*

The Story of King Krishna

May the God of History forgive me, but when I teach the decimal system in
the second or third grade I tell a story of a king, named Krishna (in honor of
the Indians), who loved gold more than anything else. Krishna had a cellar
full of gold coins. Every morning he would go down to the cellar to see how
many coins he had. But there were so many that it was difficult to estimate
their number. Therefore he asked his curator to collect the coins in bags —
10 coins to a bag. Soon enough it became hard to keep count of the bags.
So, he asked his curator to collect every 10 bags of coins in a sack. When
there were too many sacks, he collected every 10 sacks in a wooden chest.

Krishna also liked to show off his wealth. Every morning he would mark
the number of gold coins in his possession above the palace gate. On the first
day, when he had only begun his collection, he had 7 coins. He wrote above
the palace gate:

7 ⊙

On the second day, he received 3 more coins. What did he do? He
gathered all his coins in a bag of ten and wrote above the gate:

1 ℧

The following day, he received 5 more coins. Now he had a bag and 5
additional coins. He wrote:

1 ℧ 5 ⊙

Why did he draw the bag on the left and the coins on the right? Because
arithmetic, like English, is written from left to right, and the bag, containing
more coins, is more important.

The next day, he had 8 additional coins. He grouped 5 of them with the
5 single coins in an additional bag. He now had two bags, and 3 single coins.
Above the gate he wrote:

2 ℧ 3 ⊙

He continued to do the same each day, until one day his advisor asked:
"Why do you bother to draw coins and bags? Everyone already knows that

the right digit represents coins and the left represents bags. You can write only the digits!" And so, instead of writing 2 ♉ 3 ☉, the king wrote:

$$2 \quad 3$$

And that is how the decimal system was invented (at least, according to this story).

The Digit 0

On the fourth day, as mentioned, Krishna had 23 coins. On the fifth day, he had 7 more. He grouped the 3 and the 7 into one ten, and so had 3 bags, namely, 3 tens. Proudly, he wrote above the palace gate:

$$3$$

He remembered there was no need to write 3 of what. Everyone understood anyhow! But when he took a second look, he was horrified: How would they know that it didn't mean "3 coins?" He must somehow make it clear that there were 3 bags (that is, three tens)! Something must take the place of the coins, so that it would be obvious that the 3 indicates bags!

And so King Krishna invented (this time, it is the God of Arithmetic I must beg to forgive me for the literary liberty) the digit 0. He wrote 0 in the rightmost place, to indicate that there were no single coins, but that there is a place in the number for coins. And so, he wrote:

$$3 \quad 0$$

as we are used to writing the number nowadays.

The digit 0 is like the backpack a child places on a chair to declare that the seat is taken. In the number 201, it means: "There are no tens. But there is a place for tens." This is how we know that the 2 indicates hundreds. If the 0 had been omitted, the number would have been 21, where the 2 indicates 2 tens.

Historical Note

The historical truth is that the digit 0, like the principles of grouping and the place value system, was also invented by the Babylonians. However, they did not use it at the end of a number, only in the middle. The difference between 3 and 30 was deduced from context.

Another Advantage of the Decimal System — the Ease of Estimation

Decimal writing enables us to appraise a number with one glance. A brief look at the number 34,522 tells us that it contains 5 digits, and therefore it is in the tens of thousands. The first digit is 3 therefore it is approximately 30,000, like the population of a town. The additional digits, we know, are less significant than the first one.

Store owners take advantage of our habits of estimation. They price a toy at $29.99, knowing that we will pay more attention to the first digit than to the following ones. There is also a psychological aspect: The missing cent is supposedly a discount, and discounts, as is well known, have an attraction beyond their actual value.

What is Learned?

What is Learned in Elementary School?

What mathematical baggage should a child carry out of school? This was one of the first questions I asked myself when I started teaching in elementary school. I had no idea how simple the answer was: a deep understanding of the essence of the number and the four arithmetical operations.

However, this simplicity is misleading. We have just learned that behind the innocent term "the four arithmetical operations" lay two basic principles that are not simple at all: the *meaning* of the operations and the way to *calculate* them. The meaning, as mentioned, is the link to reality. Calculation, on the other hand, means figuring out the decimal representation of the result. Therefore, mastering it requires an in-depth understanding of the decimal system.

From now on say: In elementary school we learn *the meaning of the operations* and the rules derived from it, and *the decimal system.*

Fundamental Structures of Thought

A child does not enter school a blank page. He already knows, or should know, many things. As in life, the important principles are learned at an early age. And as in life, it is the basic principles that are the hardest to pinpoint. We are unaware of most of our basic mechanisms of thought. Everyone knows, for example, that if you climb up 4 steps, you have to climb down 4 steps to get back to the starting point. However, we weren't born with this knowledge. Acquiring it was a true accomplishment.

The following is a list of several structures of thought of which a child starting school has a certain perception. One should remember that the child's grasp of these principles is usually vague and intuitive, and therefore they have to be taught again, this time explicitly.

Left–Right
Up–Down
Large–Small
Before–After (in space)
Before–After (in time)
Equality (of shapes or numbers)
Symmetry

Counting Objects
Enumeration
(that is, the ability to repeat the numbers in their correct order)
Reversal
(If you're bigger than me, then I'm smaller than you. If I climbed up 3
steps, to return, I need to climb down 3 steps.)
Grouping Sets

Quite surprisingly, most items on this list deal with relations. Even "left–right" is a relationship: One object can be to the right or to the left of another.

Tip for Parents:

An important asset for a child entering school is knowledge of left–right directions. This is the preamble to the concept of order, and it is essential in all fields of mathematics. Numbers are usually written from left to right, and this is the basis for representing numbers using the decimal system.

In general, the main concepts a child should be familiar with when preparing for the first grade are concepts of relation: "before–after," "up–down," "more–less." These concepts are encountered in any game the child plays, and all a parent needs to do is draw his attention to them. When throwing dice ask: Who got more? How much more?

The Curriculum in a Nutshell

The following abridged curriculum may help you understand where your child currently stands, where he's headed, and what to expect. It is important to remember that different textbooks cover a different extent of material, and different schools deviate from the dictated curriculum in various ways. Therefore, the following is an average of kinds. The curriculum is written for Grades 1 to 6, the years covered by this book. I also added some points, derived from my personal opinion that early exposure to concepts is always beneficial. It allows the child periods of incubation between encounters with concepts. For example, first-graders should be familiar, at a primary level, with the concept of the fraction and its relationship to division. The decimal fraction can be introduced in the second and third grade, through money.

First-grade teaching begins with spatial orientation: right, left, up and down. The children are then taught the concept of the number, counting, the large-small relationship, and the order between numbers.

Later, they are taught the meaning of arithmetical operations and the decimal system. As to the meaning of the operations, the Israeli curriculum includes the meaning of addition, subtraction and multiplication (there isn't always enough time for multiplication). Slightly more ambitious curricula also include division, as was the case in Israel during the 1960s. When learning the decimal system, the child should experiment in grouping tens, and understand that more than one ten can be grouped: 30, for instance, is 3 tens. A first-grader can understand the meaning of decimal writing, and the role of digits in a number: 23 means 2 tens and 3 single ones. As to the calculation of operations, children should know how to add, subtract and multiply within 20 (that is, the result should not exceed 20). Ambitious curricula include addition and subtraction up to 100. If division is also taught, there is no need to dwell on its calculation. Fractions, such as a half, a quarter and even a third, should be taught on an introductory level. Basic measurements of time and of length should be encountered. Familiarity with time is important not only because of its practical value, but also because of the connection to fractions, and because the relationship of hours to minutes is similar to the relationship of tens to ones in the decimal system.

In the second grade, the children develop a deeper understanding of the decimal system. They are taught numbers up to 100, and calculate addition and subtraction within the boundary of 100. Ambitious curricula reach 1000. As to multiplication, in my experience it is not hard to teach the entire multiplication table towards the end of the year. Some programs make do with the multiplication table up to multiples of 6. To these I would add an introduction to the two different meanings of division (see chapter on division) and the concept of remainders in division, as well as an introduction to fractions and their relationship to division. It is important to introduce at this stage, a fraction of a group — what is $\frac{1}{2}$ of a class of 30 students? Another topic which should be introduced in this grade is measurements (pounds and ounces, feet and yards).

The third grade is mostly dedicated to the decimal system. The children are taught methods of addition and subtraction within 1000 (at least), and vertical multiplication. The concept of the fraction should be expanded. In addition, the method of calculation of division should be taught. Time measurement is learned in more depth, and the students should know how to convert minutes to hours and vice versa, including fractions of hours. The concept of volume should be taught in a concrete manner, by examining the capacity of various containers.

In the fourth grade, the children are taught calculations of addition, subtraction and multiplication in large numbers (for instance, up to a million), which require an abstract understanding of the principles studied so far, since in large numbers, it is more difficult to rely on intuition. The calculation of division ("long division") should be taught. The concept of the fraction is taught in depth. As to measurements, one reaches the relatively abstract concept of area: What a square centimeter and a square meter are.

The fifth grade is dedicated mostly to simple fractions, including operations with them, and to ratios. Factorization of numbers and the power operation are taught, both being necessary for finding a common denominator. Another subject related to fractions is mixed numbers. In some curricula, decimal fractions are taught.

In the sixth grade, the children study the decimal fractions and learn to perform operations with them. Additional subjects include ratio problems and percentages. Negative numbers can also be taught in the sixth grade, depending on the time available and any introduction that may have been made in previous years. (This book does not include negative numbers, since in most cases they are not taught.)

It is important to remember that the sixth grade is between a rock and a hard place. On one hand, it includes any material postponed from previous years. On the other hand, the last semester is affected by the imminent graduation. Therefore, the amount of material covered may be limited in comparison with the fourth and fifth grades.

The Special Role of Division

Careful scrutiny of the curriculum detailed above, and of any other curriculum in the world, reveals a surprising fact: Division has a special status. It is awarded with a greater portion of teaching time than any other operation. The turnabout occurs around the middle of the fourth grade. From this point on, until the end of the sixth grade, the children are taught the meanings of division, ratio problems (which are expressed by division) and the efficient, systematic tool used for discussing division and ratios — the fraction.

Why is division so special? Because the operations of addition and subtraction are too simple to describe the world. When things get complicated, multiplication and division are required. A large part of our world operates according to the principles of proportionality. In elections, for example, the number of mandates each party receives is more or less proportional to the number of votes it received. Proportionality is a guiding principle in understanding our environment, and proportionality is expressed by division.

Another reason for spending more time on division is that it is more difficult than the other operations. Of the four operations, it has the most meanings, it is the hardest to calculate, and the problems it can represent are the most complicated.

Spiraling

Pooh was getting rather tired of that sand-pit, and suspected it of following them about, because whichever direction they started in, they always ended up at it, and each time, as it came through the mist at them, Rabbit said triumphantly, "Now I know where we are!" and Pooh said sadly, "So do I."

A.A. Milne, ***The House on Pooh Corner***

Like the characters in *The House on Pooh Corner*, when studying arithmetic the same point is repeated over and over. But, unlike in the story, each time we are wiser and view the matter from a different angle. Educators refer to this as "spiral learning." Like a spiral, we pass over the same point again and again. And like a spiral, each time it is at a higher level.

Take, for example, the decimal system. From the vague perception of grouping tens in kindergarten, the first-grader moves on to explicit wording of the principle of grouping. He knows how to calculate $8 + 5$, by grouping a ten from the result and leaving the remaining 3 as ones. Later on he can use the exercise $8 + 5 = 13$ to calculate $28 + 5$, which contains many additional components: breaking 28 into 20 and 8, adding the 8 and the 5 to obtain 13, and joining the components back together as $13 + 20$. A third- and fourth-grader can generalize the same principle to group tens into hundreds, and hundreds into thousands. In the sixth grade, a bridge is built between fractions and the decimal system, in the form of decimal fractions and percentages.

Part 2

The Road to Abstraction — Principles of Teaching

If you can't explain a concept to a six-year-old,
you don't fully understand it.

Albert Einstein

Conveying Abstractions

The King's Road to Abstraction

During a sabbatical I spent in North America, I slipped on some ice and injured my shoulder. The doctors deliberated whether to operate or not. I consulted a renowned specialist, who was all for operating. I asked him who would perform the procedure and discovered, of course, that he would do it himself. At that moment, a piece of advice from Jackson Brown's *Life's Little Instruction Book* came to my mind: "Don't ask the barber whether you need a haircut." I decided to avoid the operation, which turned out to be the right decision.

I later thought: What would have happened if Brown had phrased his advice abstractly, like "Don't ask advice of someone with a personal interest in the matter?" I probably would not have been able to summon it at the right moment. It would not have been alive enough in my mind to be associated with the current situation. The power of Brown's advice was in that it was given through an example. It is easier to associate one specific example of a principle with another specific example than to remember the abstract principle.

If you want to convey an abstraction, use the concrete, namely an example. That is the king's road to understanding. No one can become Chief of Staff without first being a soldier and going through all the stages on the way to the top. The same is true of abstractions — you can't just place them on someone. They need to be established from the bottom up, from their concrete foundations. A piece of information can be conveyed to another person; abstractions must be established alone, and the teacher can only help by guiding to their specific examples. No person, child or adult, can understand an abstraction just from its wording. If one does understand, it is because he already had an available example in his mind.

The same is true not only of teaching mathematical ideas, but also of the way they are formed. Upon being asked how he solves his problems, Karl Friedrich Gauss (who was mentioned in the chapter *The Three Mathematical Ways of Economy*), answered: "by systematic, palpable experimentation." To solve a problem, you examine examples.

The art most familiar with the secret of conveying abstractions is poetry. Just like good teaching, a poem conveys its message through the concrete.

Instead of saying "These are the small things that make the world; our mind makes of them what it needs to make," Emily Dickinson writes:

> *To make a prairie it takes a clover and one bee, —*
> *One clover, and a bee,*
> *And revery*

Begin with the Familiar

One of the fundamental principles of teaching is to begin with the familiar. If a student already has a structure of thought in his mind, use it. This is another tangential point between teaching and poetry — the use of metaphors. The secret to the success of a poetic metaphor is that it serves a double role: It simultaneously conveys and hides information. On one hand, it is a clever scheme for conveying condensed information. On the other hand, it distracts the mind from the message, for it is supposedly about something entirely different. In teaching, the power of the metaphor lies in the first of these two roles, namely in conveying information. An entire pattern, charged with many meanings, is conveyed in one go. It is as if just as you are about to build a complicated structure in one place you suddenly discover that such a structure already exists in another place, and all you need to do is transfer it, as one piece, to where it is needed. Understanding a metaphor requires less effort on the part of the receiver, since the structure already exists in his mind.

The principle of the common denominator, for instance, can be likened to languages: In order to add fractions they must "speak the same language." For example, it is easy to add $\frac{1}{5}$ and $\frac{2}{5}$ since they both speak "the language of fifths." But how are $\frac{1}{2}$ and $\frac{2}{3}$ added? A common language must be found, like an Israeli and a Frenchman who do not understand each other's language and discover that they both understand English. In the case of $\frac{1}{2}$ and $\frac{2}{3}$, the common language is the language of sixths. Both fractions can be expressed as sixths: $\frac{1}{2} = \frac{3}{6}$ and $\frac{1}{3} = \frac{2}{6}$.

Diversity and Fixation

The Need for Diverse Examples

How do you teach a child what is a "dog"? Not by explaining, of course, but by example. However, when you point to a dog and say to the child "dog," how will he know that you don't mean only that particular dog? Or a certain type of dog? More than one example is needed in order to teach the general concept, and more than one type of example. The child needs to see big dogs and small ones, black dogs and white ones, poodles and labradors, so that he will understand that "dog" refers to the common denominator of all these. In other words, diversity is required. Lack of diversity leads to fixation, adherence to an insignificant detail that is not really associated with the abstract concept.

This is an important principle of teaching abstract concepts, particularly in mathematics. A child who counts only apples will eventually associate the number 4 with "4 apples." A child who learns to count objects that are arranged exclusively in a certain pattern will associate the number with that pattern. To understand the concept of the number, he must count various kinds of objects, arranged in different ways.

> The examples of a new concept must be diverse enough to avoid fixation on incidental details.

Example: The Concept of the "Whole"

I encountered an amusing example one day when observing a lesson on fractions at a school in Tel Aviv. Four children were given cards with the fractions $\frac{1}{5}, \frac{2}{5}, \dots, \frac{7}{5}$ and were asked to sort them according to a characteristic of their choice. It seemed that they knew what they were expected to do: They partitioned the fractions to those which are smaller than 1, equal to 1 (namely $\frac{5}{5}$) and bigger than 1. Then I asked one of them what is a fifth. He mumbled something about circles, and I understood there was a problem. I gave him a sheet of paper and asked him to show me a fifth of it. Under no circumstances could he manage. He succeeded in dividing it into 4 equal parts by drawing two lines, halving the page vertically and horizontally. But he was unable to divide it into five parts, or, when I asked him to do so, three parts, even when his three friends came to

51

his aid. All four tried to "round the rectangle," dividing the page as one would divide a circle: by lines originating in the center. The problem is that a circle can be divided into five parts in such a manner; a rectangle cannot. With rectangles the partitioning is much simpler — using 4 parallel lines.

Eventually, I told the child whom I had first addressed: I know what your problem is. The page is too big. Here is a smaller rectangle (I drew a rectangle the size of a rectangular slice of cake). This is a piece of cake, and you are now a mother who wants to divide this piece between her three children. How do you do that? He succeeded immediately — by drawing two vertical lines. From then on, the path to other concepts of fractions was easy.

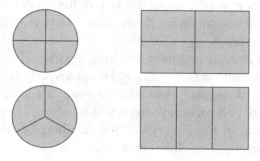

When the lesson ended, I looked at the class textbook. Lo and behold, I found that on 90 consecutive pages fractions were presented in one way only: by dividing circles, all of the same size, into "slices of pizza." Pizza slices are indeed a good example of the concept of fractions. But adhering exclusively to this model causes the children to confuse it with the abstract principle. Namely, it induces fixation.

A good textbook does the opposite. It presents the fraction, right from the start, as a part of all sorts of wholes: part of a square, a rectangle, a small circle, a large circle, and part of a group. This makes it easier to teach and learn: What is half of a group of 30 children? How many are a quarter of 20 apples? It clarifies to the child the abstractness of the notion of the "whole," namely that it can be something general.

Why Teaching is Difficult

I've been giving this lecture to first-year classes for over twenty-five years. You'd think they would begin to understand it by now.

John Littlewood, Mathematician

"You Can't Miss It"

When I first visited the United States I was surprised to realize that whenever I asked for directions the person who gave the instructions ended with "You can't miss it." In most cases, I missed. If so, what do people mean by "you can't miss it"? Only this: "I can't understand the problem. As someone who is familiar with the way, I don't see why it's difficult."

A person giving directions has lots of little pieces of knowledge to guide him: a bush at one point, a trash can at another, a road sign at a third. He has never put words to all these, and therefore he does not convey them. Unheedingly, he assumes that they also exist in the mind of the receiver.

For those who know, it is difficult to understand what others do not understand. Therefore, one should always keep in mind the ancient Hebrew sages' saying: "The shy cannot learn nor the impatient teach." Both teacher and student must be aware of the fact that the teacher knows something the student does not, that it is in both of their interests to bridge the gap, and that they both need to make a common effort to do so.

Mathematical Subtleties

The bush, the trash can and the road sign exist in mathematics as well. In mathematics, these are the small subtleties that are so easily missed. Here is an example of such a subtlety: the various meanings of subtraction.

During a first-grade lesson I observed, the children looked at a drawing of three green apples and two red ones. They were supposed to tell "arithmetical stories" based on the drawing, one of addition and one of subtraction. (We will see the importance of such "stories" when we discuss the meaning of arithmetical operations). They had no difficulty with the first task. "I had 3 green apples and 2 red apples. How many did I have altogether?" When they came to the subtraction story, confusion prevailed and as usually happens in elementary school, it manifested as inattention. Eventually one of the children said: "I had 5 apples. I ate 2. How many do I have left?"

It is easy to tell an addition story about this picture.
Can you also tell a subtraction story?

The problem was that this wasn't the "correct" story. It wasn't based on the drawing. The drawing doesn't show two apples disappearing, by being eaten or in any other way. That is why the children found the task difficult.

The difficulty is derived from a subtlety: Subtraction has more than one meaning. There is the meaning of "take away," where objects are removed: I had 5 balloons, 2 of them popped, how many do I have left? This is the meaning the child used in his story — his apples disappeared. But there is also the meaning of "whole part," where nothing disappears. There are 5 children in a group and 2 of them are boys. How many girls are there? Here, too, the exercise is 5–2, but the meaning is different. This is the meaning depicted in the drawing. The story that fits the drawing is: "I have 5 apples, 2 of them are red. How many of them are green?" To avoid confusion and anxiety, this should be clarified for the children.

Children Think Differently

An additional obstacle to recognizing missing layers is that it is difficult for adults to grasp a simple truth: Children think differently. As the poet Theodore Roethke is quoted in the introduction to this book, "A poet must have his childhood close at hand." A poet must be able to connect mainly with his feelings. A good teacher needs to connect with the way the child within her thinks.

It is not easy at all. Patterns of thought change with age, and as we grow older we tend to forget them, in mathematics as in all other areas of life.

Mediation

A mathematics teacher is midwife to ideas.

George Pólya, Mathematician

It Must Arise from Within the Child

My two-year-old nephew sat on a chair belonging to his four-year-old brother, Rotem. He was very pleased with himself, but his brother, of course, was less happy. The parents tried to convince the younger child — the chair isn't yours, you have your own chair, each child has his own chair. But all their imploring and persuading were to no avail. Then, I turned to him and asked: "Whose chair is this?" "Rotem's," he said and got off.

Only later did I notice that I learned this trick in elementary school. Educators have given it a special name: "mediation," meaning to cause the solution to stem from within the child. When something stems from within, its understanding is completely different from something that is heard from others. This is true of teaching in general, and particularly of teaching young children. Anybody who has taught in elementary school knows that children do not passively absorb facts. To understand why $3 + 2 = 5$, for instance, the child must do more than hear this from his teacher — he must experience it himself.

If so, what is the teacher's role? Why is she necessary?

What does "Teaching" Mean?

I have been teaching for 25 years, and I still wonder: What does "teaching" mean? What does a teacher do in a class?

The answer is seemingly simple: To teach is to transfer knowledge from someone who has it to someone who does not. What can be simpler? But what does "to transfer" mean? And what is the student's role in this transfer? Does he remain passive? Is his only job to wait for the teacher to pour the knowledge into his mind? Establishing concepts is not a transfusion of knowledge, just as building a house requires more than just spilling concrete and iron onto a building site. The teacher and the student need to work together to create the concepts in the student's mind.

The teacher's advantage lies in the fact that she knows where they're going. Her role is to guide the student's attention to the correct notions in the proper order.

Example of Mediating in the First Grade

Here is an example of mediation in the first grade.

Teacher: Which number is greater, 7 or 5?
Students: 7.
Teacher: Which of the numbers 3, 5 and 6 is the greatest?
Students: 6.
Teacher: What is the difference between the two questions I asked?
Students: In the first question you asked which number was greater. In the second you asked which was greatest.
Teacher: That's right, or in other words, which is the biggest. When do you ask which is greater and when which is greatest?
Students: When there are two numbers you ask which is greater. When there are three numbers you ask which is greatest.
Teacher: Do there have to be three numbers?
Students: No. There can be more than that.
Teacher: Give me an example of the question "which is the greatest number" with more than three numbers.
Students: Which of the numbers 4, 10, 5 and 13 is the greatest?
Teacher: Very good.

The above mediating process not only introduced the concepts of "greater" and "greatest," it also accomplished at least two other goals. First — leading the children to reflection, or introspection: What did we do? What is it that we were asked? What was the difference between the wordings of the questions? Another aim achieved was leading the children to find examples themselves. The answers were provided by the children; the guidance, as well as the accurately worded concepts, using the proper language, was provided by the teacher.

Even the Obvious Must be Stated

Have you ever disliked a lecture because it was too elementary?
                                    Paul Halmos, Mathematician,
                             from ***How to Talk Mathematics***

"Give me an example of the question *which is the greatest number* with more than three numbers," asked the teacher of her students in the dialog above. This is not a difficult task. The teacher had no doubt of the children's ability to perform it. And yet, it was valuable.

"From within the child" does not mean putting the child to a test, leaving him to struggle on his own and imposing difficult tasks on him. It means gently guiding him towards the correct concepts. Even the simplest things must be experienced, and even the obvious needs to be stated.

For the last three decades, a trend in education, "constructivism," has interpreted the "from within the child" approach on the extreme side: The child is supposed to discover the principles and rules on his own. This is not mediation. Mediation is not about self-discovery of concepts, but about joint experiencing of that, which the child is able to achieve.

Another Example of Mediation

I borrowed the following example from a teacher friend of mine. I am particularly fond of it, and will explain why later:

Teacher: Open your new book to Page 6. Why are we starting on Page 6 and not on Page 1?
Student: The book has a cover page with the name of the book and the grade. That's Page 1. Then there's an internal cover page, that's Page 2. Page 3 is the Introduction. Then there's a Table of Contents on Pages 4 and 5. The real text only starts on Page 6.
Teacher: Good. What do we look at when we begin a new chapter?
Student: The title.
Teacher: Why?
Student: The title helps us understand what we are about to study.

What is so nice about this dialog? First of all, most teachers would have skipped this stage (I most certainly would have). Such omission leaves the child with a sort of blank, an area that does not receive recognition — what happens up to Page 6? Second, the dialog encouraged the useful habit of reading introductions and titles and paying attention to them.

It is Easier to Talk Than to Listen

In after-years he liked to think that he had been in Very Great Danger during the Terrible Flood, but the only danger he had really been in was the last

half-hour of his imprisonment, when Owl, who had just flown up, sat on a
branch of his tree to comfort him, and told him a very long story about an
aunt who had once laid a seagull's egg by mistake, and the story went on
and on, rather like this sentence...

A.A. Milne, **Winnie the Pooh**

It is a strange fact of life that it is much easier to talk than to listen.
A lecturer can talk for hours. If he doesn't engage his audience actively,
they will grow tired after a short while. Listening is not as passive as it
seems. Understanding requires effort and active digestion of the concepts.
An ongoing lecture does not provide the leisure required for such absorption.
Studies have shown that a university student becomes inattentive after, at
most, fifteen minutes of hearing a lecture. Then his thoughts begin to wander.

If this is true of university students, it is most certainly true of elementary
school students. And when a young child becomes inattentive, it is much
easier to notice. Even in the university it is necessary to pause once in a
while and conduct a joint discussion. In elementary school, class discussions
and active participation are the only ways to teach.

Magic Words

What's in a name? That which we call a rose by any other name would smell as sweet.

Shakespeare, ***Romeo and Juliet***

What is in a name? A whole lot. It is not what you call something, a rose or any other name, but the fact that you name it. A name identifies a pattern in the world, and indicates that it exists. There is good reason in that the first thing Adam did was name the animals.

How Important is Naming in Elementary School?

When I began teaching in elementary school, I was convinced that children should not be burdened with too many concepts. They should be taught through examples. Explicit wording is for adults only. One of the surprises that awaited me was discovering how wrong I was on this particular point. Children love terms, names and explicit wording. They are proud of their ability to use them. I remember one time when a second-grade teacher unheedingly said, "There are other options here." Although she immediately corrected herself: "other possibilities," not a moment had passed before one of the children waved his hand excitedly and called out, "There is another option! There is another option!" In one of my first-grade classes, I used the expression "et cetera." This caused a great deal of excitement: The children repeated it over and over, just like the King of Siam in the musical *The King and I*, who proudly used the exact same expression. Even more than with names, children are proud of their mastery of notation. Nobody will be prouder than a first-grader who has learned the notation of a "half," and is able to explain the meaning of the 2 in the denominator.

Thoughts do not originate in words, but words are the scaffoldings that allow the building of high towers of ideas. Ideas are first formed intuitively, but once formed, if they are to be used as the basis for the construction of further concepts, they have to be consolidated and formulated precisely. Naming is not to be confused with ideas and definitions are not to be confused with theorems. But in order to enable thinking and communicating about mathematics, names should be given. So, when learning division, the children should learn to use the words "dividend" (the number divided), "divisor" (the number dividing) and "quotient" (the result of the division).

This will enable them to discuss with precision the operation of division and, later, fractions.

A First-Grade Example

I was lucky enough to have the opportunity to participate in the teaching of the same page in a certain textbook in three different classes within the span of one week. This enabled me to try various approaches and compare their effectiveness.

I mentioned this page before, in the chapter *Why Teaching is Difficult* — it includes a drawing of 3 green apples and 2 red ones, and the children are asked to tell addition and subtraction stories. I described one lesson where the children's attention had not been drawn to the various meanings of subtraction which resulted in confusion.

The following week, another class in the same school reached that same page. By now experienced, I stopped the lesson and said to the children: Before we begin this page, please tell me a story of subtraction. As I expected, the example they gave me was of the "take away" type, that is, a story where things disappear. Something like "I had 5 balloons, 2 of them popped, how many are left?" Good, I said to the children. But you should know that there are subtraction stories in which nothing disappears. Instead, they include different kinds of objects. We know how many there are of one kind, and ask how many there are of another. For example, "There are 5 children in a group. 2 of them are girls. How many boys are there?" The answer is a result of subtraction: 5–2. In this case, we sort according to gender.

Then we moved on to the apples and their arithmetical story. This time, the children had no problem: I have five apples, 2 of them are red, how many are green? In this case, by the way, we sort according to color.

In the third class, I tried a different approach. Once again I stopped the lesson, only this time I didn't use explicit wording. Instead, I used an example: Before we begin the story, I said, I will tell you a similar story. And I told them the story of the 5 children, of which 2 are girls.

This trick didn't work. The children still found it difficult. My explanation wasn't explicit enough, and the children did not have enough time to take in the principle. For me, this lesson was extremely instructive, showing me that exact naming should not be skipped. Presenting matters in an explicit way, clearly defining the concepts, always helps.

By the way, I would do things differently today: I would draw the formulation of the principles from the children. I would start with the example of the 5 children, 2 of whom are girls, and generate a discussion in class on this strange type of subtraction in which nothing disappears. The principles should come from the child.

The Importance of Clear Instructions

Along with clear naming, there is an additional principle in teaching, more important in elementary school than in any other place: The teacher's instructions must be clear. I have witnessed unclear instructions in countless lessons, and they always lead to the same result: Instead of trying to solve the problem, the children try to aim for the teacher's mind, guessing her intention. The message conveyed is that the teacher's thoughts are more important than their own, whereas the message should be: "We are in this together to discover the rules that govern the world."

Guessing

A famous Hungarian teacher insists on guessing before calculating. Before the student calculates the precise result of 654×321, he or she should try to estimate the result. Is it more than a million, or less? (It is less, since 654 is smaller than 1000, and so is 321, so the result is less than 1000×1000). Is it more than 100,000 or less? (It is more, since it is larger than 600×300, which is 180,000.)

Why is it so important? A common answer is — getting a feeling for numbers. Thinking first in terms of order of magnitude gives a sense of "large" and "small". Guessing also encourages experimentation, testing your conjectures against the data.

For example: Joe's brother is twice as old as Joe, and the sum of their ages is 18. How old is Joe?

This may be difficult for students of fourth or fifth grades. But ask them to guess, and they will promptly reach the answer. They will experiment: can it be more than 9? No, because twice of 9 is 18, and in the problem 18 is more than twice Joe's age. Trying 5 will yield the sum 15, so it should be more than 5. Experimentation will lead the students to understand that the sum of the ages of the brothers is in fact 3 times Joe's age, which easily leads to the correct answer: 6.

But probably the most important benefit of guessing is that it relieves anxiety. Guessing means permission to make mistakes. There is no one correct answer; getting close is good enough.

And finally guessing is fun. It is a bit like playing darts, and trying to hit the mark.

The Calculator and Other Aids

Should Calculators be Introduced into Elementary Schools?

Over the last few decades, there is growing usage of the calculator in the elementary school. The damage is immense. I meet high school kids who use a calculator to calculate 3 times 15. I have even heard of a student in a prestigious department at my university who asked for a calculator during an exam to calculate 7 times 8.

What's wrong with this? Calculations contain principles, and principles are assimilated only through practice. Take, for example, the calculation of 3 times 15. It contains two principles. The first is that: $10 + 5 = 15$. This sounds obvious, but it isn't. "Know from where you have come" is as important in numbers as it is in life, and the representation of numbers comes from the decimal system. When breaking 15 into a 10 and a 5, one returns to the principles of the decimal system.

The second principle is the distributive law: When 3 is multiplied by $10 + 5$, it is multiplied both by 10 and by 5 since 3 times $10 + 5$ means: $(10+5) + (10+5) + (10+5)$, which is 3 times 10 plus 3 times 5. This insight is lost when using a calculator.

One of the arguments in favor of the use of the calculator is that it enables children who missed a certain stage to make up for the loss. For example, a child who did not learn how to multiply and divide can use a calculator to solve more complex problems, such as ratio problems. The necessary multiplication and division is performed for him by the calculator.

Such an argument demonstrates a basic lack of understanding. No aid can constitute a king's road to mathematics, most certainly not the calculator. There is no skipping in mathematics. A child who did not learn the basic calculations cannot move on to advanced mathematics. A bypass in mathematics can only lead to one place: mathematical ignorance, accompanied by mathematics anxiety.

The harm of the calculator was demonstrated in the international mathematics tests held every few years. In an analysis of the results of the tests it was found that in the 5 countries which achieved the first places in those tests, the use of calculators in the elementary school is extremely limited. In the 10 countries placed last, the use of calculators is much more extensive than in the 10 countries placed first.

Additional Aids

The calculator is the ultimate aid. No understanding is required to use it. But aside from it, there are many other types of aids: various illustrations of the decimal system, which group tens for the child; aids that divide circles into equal parts and slides to help with expansion of fractions. There is no limit to the inventors' imagination.

All these are like bringing cars to PE class to save the children the effort of running. No one can perform the task of understanding for the child. Each person must undergo the process himself, experiment with the concrete basis of abstractions on his own. Without such experimentation, understanding is verbal and perfunctory.

Mental Calculations

Even pencils and paper are aids. They cannot be completely avoided because of the limitation of our memory. But when it is possible to forgo them, it is worthwhile. One of the best ways to understand the decimal system is to require the children to perform "mental calculations," namely, to calculate in their heads. For example, multiplying a two-digit number by a single-digit number (for example, 17×8). On paper, this type of exercise can be performed automatically. To calculate it in his head, the child must understand the decimal structure and practice the decimal system. The importance of mental calculations has been shown in the international mathematics tests: Those countries that encourage mental calculations attain higher places than those that do not.

Patterns as Aids

A special type of aid is patterns, into which the child is supposed to enter data and receive the result in some mechanical way. The most famous of these is "the rule of three," for solving ratio problems, dreaded by generations of students. This is a pattern in which numbers are placed and then multiplied and divided according to some rule which few people manage to remember over time. Teachers who use it claim that this is the only way to bring some children to solve ratio problems. Such an argument should be regarded much the same as the arguments concerning the advantages of the calculator: Aids do not lead to understanding. It is better that a child not learn how to solve

"rule of three" problems than he follow a rule he does not understand, a sure-fire way to create mathematics anxiety.

In Praise of the Personal Board

There is one teaching instrument that is not an "aid," but a tool much like a notebook and pencil. Discovering it was a turning point in teaching for me, and I would therefore like to recommend it: the personal board, a descendant of the once-used slate. It is a board, the size of a big page or slightly larger, on which the child can write or draw and then lift up to show his teacher. There are white boards using markers and boards using chalk, and some improvise with laminated paper and erasable markers. The technical aspect isn't important. In all the classes in which I observed the use of such a board it was an extremely useful tool.

There are several reasons. First, the board provides the child with the sense of pride of being a little teacher: He is writing like his teacher. Second, it enables writing in large handwriting, which is easier for children in the lower grades. But most importantly, when lifted, the entire class can see it, inspiring a sense of "we're all together on one mission," a feeling that writing in notebooks does not provide. Also, the teacher can scan the products of all her students in an instant and learn quickly about any problems in their understanding.

A small tip on the use of the board: Having completed a task, the student should not raise his board immediately. He should turn it over, as a sign that he has done his work, and only when the teacher signals, the entire class should raise their boards together. This way the spirit of competition is avoided, and the sense of togetherness is enhanced.

The Courage for Simplicity

We are about to move on from general principles to the mathematical material itself. Can we take with us, as food for the journey, a short summary of all we have learned so far?

The following is an attempt at such a summary. The first principle is that the secret of proper teaching of arithmetic lies not in didactics but in being familiar with mathematics itself. In particular, understanding the fine layers of which the concepts consist.

The second principle can be termed "courage for simplicity." The teacher's role is to enable the child to directly experiment with the mathematical principles. This means also experimenting with the simplest things: The secret lies not in sophistication but in establishing concepts from their foundation. In addition, it is important to use explicit and accurate wording. In this, too, the simple should not be feared. Even the most obvious must be phrased explicitly.

The Courage for Simplicity

Part 3

Arithmetic from First to Sixth Grades

A. Meaning

The meaning of numbers and arithmetical operations is their link to reality. This is the first stage, preceding calculation. Teaching it is the first step in teaching arithmetic, and if taught properly, perhaps also the most enjoyable.

The meaning of the number is derived from counting objects. The meaning of addition is the joining of two groups. The meaning of subtraction is removal. The meaning of multiplication is repetition of equally sized groups. The meaning of division is dividing into equal parts.

One of the things that make the meaning interesting is the fact that it contains subtleties. For instance, there is a fine distinction between joining which involves adding more objects of the same kind, and joining of various types of objects that are constantly there. A similar subtlety of meaning exists in subtraction. Subtraction has two entirely different meanings, and it is essential to emphasize them.

The meaning also determines the rules that govern the operations.

The Meaning of Addition

Look, we are two numbers
Standing together and adding
Or subtracting, since after all the sign
Changes from time to time.

Yehuda Amichai, "Up On the Acorn Tree," ***Poems***

Addition is Joining

The expression $3 + 2$ applies to the joining of two groups, one consisting of 3 members and the other of 2 members. Joseph has 3 flowers, Reena has 2 flowers. How many flowers do they have altogether?

Seemingly, there is nothing simpler. However, before we go any further, we must discern a subtlety of meaning. There are actually two different forms of addition: dynamic and static. In dynamic addition, to join means to change a situation: 3 birds were sitting on a tree, 2 joined them. How many birds are there now? In static addition, joining signifies grouping of types: A vase contains 3 red flowers and 2 yellow flowers. How many flowers are there altogether?

Dynamic addition: 3 birds were sitting on a tree, 2 joined them

Static addition: a vase contains 3 red and 2 yellow flowers

I know a teacher who terms the first type "movie" and the second "picture." I myself don't use these terms in class, but I do emphasize the difference, especially because of the link to subtraction. There, too, such a distinction is made, and there it is essential, since children find static subtraction difficult.

The terms of an addition exercise are called **addends**. Thus, in the exercise $2 + 3 = 5$, the numbers 2 and 3 are the addends, and the result, 5, is the **sum**.

Common Denomination

3 **pencils** plus 4 **pencils** equal 7 **pencils**. The sum has the same denomination as the addends. In other words, we are adding objects of the same kind. What happens when the two addends have different denominations, as in the example: "How many are 3 bananas and 4 oranges?"

To perform this addition, a common denomination is required: "fruit," for example, or "objects." 3 bananas plus 4 oranges equal 7 fruits, or 7 objects.

The most famous example of the importance of a common denomination is found in the addition of fractions. One seventh can be added to two sevenths since they both have the same denomination. Together, they equal three sevenths. In contrast, to add one seventh and two thirds, both fractions must first be expressed using the same denominator. In other words, a "common denominator" must be found.

Some are deterred from teaching the principle of common denomination at such an early stage. In fact, the children really like it. What common denomination can you find for lions and tigers? Chairs and tables? How many are 2 lions plus 3 tigers? 3 chairs plus 4 tables?

The Commutative Law: Are $3 + 4$ and $4 + 3$ the Same?

The result is, of course, identical: 7. This is "the commutative law," its name derived from the fact that the two addends exchange places. Does this mean that the meaning is identical? In static addition, yes. The question: "George has 4 pencils, Molly has 3. How many do they have altogether?" is identical to "Molly has 3 pencils, George has 4. How many do they have altogether?" This is not true of dynamic addition: The story "4 stories were added to a 3-storey building" is not the same as "3 stories were added to a 4-storey building." But even in the second case, the difference isn't very significant. Both addends have the same denomination (in the example above both the 3 and the 4 indicate "stories"), and therefore the difference in their order does not have a deep meaning. There is no difference between the added numbers in English either — they are both called "addends."

If the difference in the meaning of $4 + 3$ and $3 + 4$ isn't significant, is it really worth devoting time to it? The answer is definitely "yes." Even if only for the reason that the children like it.

Another reason why this rule is important is that sometimes switching the order makes the calculation easier. For instance, it is easier to calculate

9 + 2 than 2 + 9: in the first you begin at 9 and take 2 steps forward, whereas in the second, beginning at 2, you need to take 9 steps forward. (Still another reason is that this is an introduction to the commutative law in multiplication.) To teach this law, I usually ask a student to hold 4 pencils in her right hand, 3 in her left hand, and raise her hands while facing the class. "What is the arithmetical story," I ask. "4 + 3," the class replies (since they see the 4 pencils on the left side, and we usually start from the left). Now I ask the student to cross her hands. What is the story now? "3 + 4," they say, and on the board we write "4 + 3 = 3 + 4."

Historical Note

What is the origin of the "+" sign for addition?

Up until about 600 years ago, arithmetical operations were written in Latin words. Addition was indicated by the word "et," meaning "and." The letter "t," because of the cross at its top, was substituted at some point by a "+" sign.

The Rules of Change

The meanings of the operations dictate the rules that govern them.

The most basic rules are the "Rules of Change." They determine what happens to the result of an operation when one of its components is altered. For instance: What happens to 7, the result of 4 + 3, if 2 is added to the 3? Or if 1 is subtracted from the 4?

In the addition operation, the Rules of Change are very simple. They are less so in subtraction. In multiplication and division these rules are particularly useful. For example, the rules of multiplication and division of fractions are derived from them. Once a child understands these rules for adding and subtracting, he will not find it difficult to understand them in multiplication and division.

The Rule of Change in Addition

What happens to the result of the exercise 4 + 3 if we increase the addend 3 by 2, that is, replace it with a 5? Instead of 7, the previous result, we now have: 4 + 5 = 9. In other words, the sum also increases by 2. Thus the rule of change in addition is:

When one of the addends is increased by a quantity, the sum is increased by the same quantity.

Though this is not obvious, the rule of change actually includes breaking down one of the addends. The example above can be written as: $4 + 5 = (4 + 3) + 2$. In other words, since the 5 can be broken into two parts, 3 and 2, it can be added in two stages: first adding the 3 and then the 2. This is often the basis for calculations. For example, when calculating $50 + 23$, we break the 23 into $20 + 3$, almost unheedingly, and then calculate: $50 + 20$, which is 70, and add the 3 to get 73.

Now we can ask ourselves the reverse question: What happens when one of the addends is decreased? Since increasing an addend increases the sum, it is obvious that the reverse is also true:

When one of the addends is decreased by a quantity, the sum is also decreased by this quantity.

Here is an example of the use of this law: How do we calculate $76 + 99$? We can easily calculate $76 + 100$ to get 176. Since 99 is less than 100 by 1, according to the rule of change $76 + 99$ is also less than $76 + 100$ by 1, that is, it is 175.

In the first grade, the addition table for 9 should be taught in this manner. One can ask: How much is $10 + 8$. 18, of course. How much is $9 + 8$? The children know: 1 less, that is, 17.

In the higher grades, $365 + 999$ can be calculated in a similar way — can you figure it out?

The rule of change in addition also has a formal name: the associative law. The origin of the name will become clear when we discuss the same rule in multiplication.

The Meaning of Subtraction

Subtraction Stories

In one of the most enjoyable first-grade lessons I ever taught, I asked the students to make up subtraction stories. The rule was that the important word, the one indicating the subtraction, cannot be repeated. A new one must be invented each time. We wrote these words on the blackboard. The children were brimming with creativity: I had 5 balloons, three of them popped, how many do I have left? (We wrote "popped" on the blackboard). I had 100 candies, I ate 90. Danny had 5 cars, 5 of them broke down. (I encouraged them to include the number 0 in the exercise or the result.) And so on: fell down, broke, disappeared, wilted, were eaten up. I had promised the children in advance that subtraction was much more interesting than addition. It arises in many more types of situations. This is due to the well-known fact that it is easier to destroy than to build...

The children competed with each other in making up exercises with large numbers. So I asked them for the smallest exercise they could come up with.

Slowly the numbers decreased, until one girl said: "0 − 0, but I don't have a story." I asked the class how many elephants they think I have at home. Zero, they answered. So today, I told them, they all ran away.

In the exercise 7 − 4 the 7 is called the **minuend** (on which the action is performed), and the 4 is called the **subtrahend** (performing the action). The result is called the **difference**.

The Three Meanings of Subtraction

In all the examples above, subtraction had one meaning: removal. Part of the group is removed, and the question is how much is left. This is dynamic subtraction, where the situation changes over time. The term used in class for this type of subtraction is "take away."

But there are at least two additional meanings to subtraction. One is termed "whole part."

In a group of 5 children, 2 are girls. How many boys are there?

The answer, of course, is 5 − 2. There are two types of objects. The total of all objects and the number of objects of one type are known, and we are asked to find the number of objects of the other type.

The third meaning of subtraction is comparison: By how many is amount A greater than amount B? For example, Joseph has 7 cats, and Reena has 4 dogs. How many more cats does Joseph have than Reena has dogs?

The Rules of Change in Subtraction

What will happen to the difference 7 − 4 if the minuend, 7, is replaced by a number greater by 2, namely, 9? Instead of 3, the result of the previous subtraction, we now have 9 − 4, that is, 5. The difference is also greater by 2. In the more advanced notation of parentheses (not yet used in Grades 1 and 2), this is written:

$$(7 + 2) - 4 = (7 - 4) + 2.$$

The first rule of change in subtraction is therefore:

When the minuend is increased by a quantity, the difference is also increased by the same quantity.

And what happens if the subtrahend, 4, is increased by 2? If we subtracted 2 more, we now have 2 less. Formally:

$$7 - (4 + 2) = (7 - 4) - 2.$$

When the <u>subtrahend</u> is increased by a quantity, the difference is <u>decreased</u> by the same quantity.

Here is a little teaching trick. When teaching this principle, don't ask: "By how much is $7 - 4$ greater than $7 - (4 + 2)$," but only: "Which of the two is greater?" Better yet, use a story to demonstrate: "Reena and Joseph have an equal number of pencils. Joseph gave away 4 of his pencils and Reena gave away 6 of her pencils. Who has more pencils now?" The child will discover on his own not only that Joseph has more pencils, but that he has 2 more. Discovering on one's own, without being explicitly asked, is invaluable.

Examples of the Use of the Rules of Change

In the previous chapter, we noted the usefulness of the rule of change in calculating addition. The same is true of subtraction. For instance, to calculate $80 - 23$, the 23 is broken into $20 + 3$. First, 20 is subtracted from 80 (the result is 60) and then 3 is subtracted (resulting in 57).

To calculate $14 - 6$, the 6 can be broken down into $4 + 2$. Therefore, according to the rules of change, in order to subtract the number 6 from 14, first 4 is subtracted (a simple operation resulting in 10) and then 2 is subtracted, resulting in 8.

Parentheses

In the last section, and even before, we used parentheses. The role of the parentheses is to indicate priorities of order in performing the operations. Their content is calculated first. For instance, $7 - (3 + 2)$ means that first the exercise $3 + 2$, contained in the parentheses, is calculated and then the sum, 5, is subtracted from the 7, resulting in $7 - 5 = 2$.

Parentheses are like a box. First the content of the box is calculated, and then it is used as a unit. Just as a box can contain additional boxes, so can parentheses contain additional parentheses. In the exercise $7 - ((3 + 2) - (1 + 1))$, first the $3 + 2$ and the $1 + 1$ are calculated, resulting in 5 and 2, respectively. Then they are placed in the external parenthesis: $(5 - 2) = 3$, and the 3 is entered back into the exercise: $7 - 3 = 4$. Parentheses, and the order of operations, are taught only from Grade 3 onwards.

The Essence of Multiplication

Multiplication is Like Counting

As mentioned before, numbers were invented for concision. Instead of saying "kiss kiss kiss..." we can say "a thousand kisses." Multiplication was invented for a similar reason: Instead of saying "2 plus 2 plus 2..." we say, in short, "a thousand times 2."

The similarity between multiplication and counting is not accidental. The two are very closely related. As in counting, the basis of multiplication is the creation of a single unit. And as in counting, in multiplication the unit is repeated several times. In this case, the unit is a set. For instance, 3 times 2 means that a set of 2 members is repeated 3 times, and the question is how many are there altogether.

3 times 2   2 times 3

Since the joining of groups is expressed by addition, the same can be written as: $2 + 2 + 2$. Thus, multiplication is the repeated addition of the same number. The above repeated addition can be marked in short as: 3×2.

Historical Note

The multiplication sign, $\times$, was invented in the 17th century. In algebra, where numbers are denoted by letters, the sign is replaced by a dot: Given two numbers x and y, the operation x times y is written $x \cdot y$. The purpose of this notation, introduced by Leibnitz (1646–1716), was to avoid confusion between the multiplication sign and the letter x. In algebra the multiplication sign is often completely omitted, and we simply write xy. There is a deep reason behind it: Multiplication is like counting, since "2 apples" and "2 times an apple" are the same. And in counting, we do not write a sign between the number and the denomination, we just say "2 apples," or draw "2" next to an apple.

Economy in Calculation

We have seen that multiplication economizes in representation. Instead of many operations, only one operation is written. But the true economy of multiplication lies in a different place: calculation. If humanity had stopped at addition, the calculation of the number of legs of the entire population of the United States would have required about 300 million operations of addition!

This economy, by the way, begins only when multiplying numbers greater than 10. Calculating 4×2 is no simpler than calculating $2 + 2 + 2 + 2$, since the only way to calculate it is to add 2 four times. In contrast, when calculating 14 times 2, there is no need to add the number 2 fourteen times: It is calculated as 4 times 2, plus 10 times 2, and 10 times 2 is simple — it is none other than 2 tens.

"Multiplied by" or "Times"?

The meaning of "times" is what gives the components of the multiplication operation their names. In "3 times 4" the 3 acts on the 4 — it indicates how many times the 4 is repeated. Therefore, it is called the **multiplier**, and the 4, on which the operation is performed, is called the **multiplicand**. By the way, this does not have a full consensus. The Encyclopedia Britannica uses these terms, but in various dictionaries the terms appear the other way around. There is room here for an official decision, and I believe that the terms should fit the term "times" — the number on the left is the multiplier and the number on the right is the multiplicand.

First Lesson on Multiplication: Counting Hands

The children's own bodies are always a good starting point. Counting hands, feet or fingers provides the most direct link to reality. At some point, the children should cut loose and leave the body and move on to other objects, but for starters there is nothing simpler than counting body parts.

I ask 5 children to come to the front of the classroom, and I announce that we will now examine how many hands they all have. One child raises his hands. How many hands does he have? Two, the children say. That's right. For now we have 1 times 2. Next the second child raises his hands: Now we have 2 times 2. Then the third child raises his: 3 times 2, and finally, when all have their hands raised: 5 times 2. We count: 5 times 2 is 10.

Can we say the same using "plus"? I ask. Indeed: $2 + 2 + 2 + 2 + 2$. In this formula the number 2 appears 5 times. It is, therefore, shorter to say: "5 times 2." But in arithmetic we like to use signs instead of words. "Times" is indicated in arithmetic by "×" and so, instead of "5 times 2" we write "5×2". Look how much shorter this expression is than $2 + 2 + 2 + 2 + 2$! Arithmetic likes to be concise.

We write on the board: $5 \times 2 = 10$. The sentence should be written down immediately after the arithmetical story. The children should write the sentences corresponding to "2 times 2, 3 times 2, 4 times 2" on their own. They will do so enthusiastically!

A Little More Abstract — Feet

The next stage is to do the same with feet. The reason is that it is more abstract! Feet cannot be seen, because they are hidden under the desk. This exercise requires mental calculation. Select a group of children sitting by a desk and ask how many feet they have altogether. I recommend choosing the same number of children as previously chosen, for example, 5 (we counted the hands of 5 children). Although the children will know the answer immediately, they will also learn something: If 5 children have 10 hands, they also have 10 feet. In arithmetic, these are not the objects that matter, but their number. Then count the number of feet of all the children in the class and write down the exercise. Show them how long it would be to write the exercise using addition: $2 + 2 + 2 + \cdots$ and how much effort we save by using multiplication.

Multiples of 10

Now is a good time to demonstrate multiples of 10. This is an introduction to the decimal system. I invite a child to the blackboard and ask him to hold up his fingers. He then writes the number of fingers on the blackboard above him — 10. He stays standing next to the number he wrote, and a second child comes up, holds up his fingers and writes how many fingers he has along with the first child on the blackboard above him — 20. The class says out loud: "2 times 10 is 20." Then a third child joins them in a row, and so forth. If possible, use the entire class. The children get very excited when they reach 100 and 200.

Now we discover what happened: What did Child Number 1 write? He wrote 1 with a 0 next to it. And Child Number 2? He wrote 2, with a zero next to it. Child Number 13 wrote 13 with a 0 next to it! We discover the rule: 13 times 10 adds a 0 to the right of the 13. Multiplying by 10 adds a 0 to the right of the number! At this point in their studies, the children cannot really understand why this is so; still they can learn the rule.

The Jumper Family

As an illustration of multiplication, I sometimes tell the children a story about a family (the children chose the name "Jumper"), with 2 children, a father, a mother and a grandfather. The younger child advances one tile with each step. The older child advances 2 tiles, the mother 3, the father 4 and the grandfather 5.

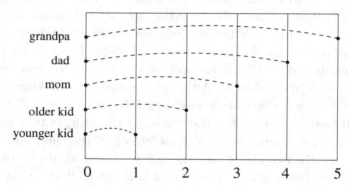

I draw a number axis with "tiles" all along it. The people stand on the lines separating the tiles.

The younger child takes 3 steps. How many tiles did he pass? 3 times 1, that is, 3.

The second child takes 3 steps. How many tiles did he pass? 3 times 2. And the father, in 3 steps? And the mother? And the grandfather? I write: 5 + 5 + 5 (for the grandfather), and say: How many times 5 do I have here? Well, how many tiles are 3 times 5? We do the same for each family member.

Again, multiples of 10 should be introduced at this stage. The family is joined by a kangaroo that advances 10 tiles with each leap. (It must be a young kangaroo — that's a small leap for a kangaroo...) How many tiles did he pass in 3 leaps?

The Commutative Law of Multiplication

We now arrive at an important issue: the commutative law, which says that **changing the places of the multiplier and the multiplicand does not change the product**. It should be emphasized in multiplication since here changing the order of the factors signifies a real change in meaning. When adding, the meaning of $3 + 2$ is not much different from the meaning of $2 + 3$. In contrast, the meaning of 2×3 and 3×2 is completely different. This is because the two factors have different roles in multiplication. Compare, for example, the drawing representing $3 + 2$:

$$||| \ ||$$

with the drawing representing 3×2:

$$|| \ || \ ||$$

In addition the two groups have a similar role. This is not so in multiplication, where the 3 stands for 3 *sets*, while the 2 for 2 *lines* in each set. They have entirely different roles! In this case, the two factors even have different denominations: "sets" and "lines." The denomination of the result in multiplication also indicates the difference between the factors: It has the same denomination as the multiplicand, not the multiplier (which does not have a denomination). For instance: 3 times 2 lines are 6 *lines*.

The difference in the status of the two factors in multiplication means that changing their order will also change their meaning. For example, 2×3 means "2 times 3," which is $3 + 3$. In contrast, 3×2 indicates "3 times 2," which is $2 + 2 + 2$. This is an entirely different arithmetical story.

Why is $3 \times 2 = 2 \times 3$?

I am sure at least some of the readers will raise their eyebrows at this question. Of course they're equal! What's to explain? Still, it is not clear at all. We have just said that the two have a different meaning!

When I come home from a day of teaching at school, I tell my children, if they are willing to listen, what I did that day. One day I told my daughter, then in the second grade, that I had taught the children why $2 \times 3 = 3 \times 2$. She was annoyed. It's the same thing! I asked her to show me with her fingers 2 times 3 (that is, 2×3). She lifted 3 fingers of one hand and 3 fingers of the other. Now show me 3 times 2 (that is, 3×2), I asked. She lifted 2 fingers of one hand, 2 fingers of the other hand, and I lifted 2 — together 3 times 2. Now show me that they are both the same, I asked.

To my surprise, without pausing to think, she lifted three fingers of her right hand, three fingers of her left hand (2 times 3), and put both hands together, so that the fingers made up three pairs: 3 times 2!

My daughter provided here a mathematical proof of the commutative law. In fact, this is just a vivid demonstration of the standard proof given for this law, which is done by looking at objects arranged in a rectangle.

When looking at the fingers from above, they are arranged as a rectangle:

$$\otimes \quad \otimes \quad \otimes$$
$$\otimes \quad \otimes \quad \otimes$$

The number of circles in the rectangle is 2 times 3, since there are 2 rows, each containing 3 circles. But it is also 3 times 2, since there are 3 columns, each containing 2 circles.

A similar rectangle, with 3 rows and 4 columns, will show us that $3 \times 4 = 4 \times 3$, and the same is true for any pair of numbers.

Axioms and Theorems

"Axiom" is a Greek word, meaning an assumption that is accepted without proof. It is used to prove other propositions, but it itself is not proved. A proposition that can be proved is called a theorem. The term "axiom" was instituted by Euclid. He founded his geometry on five axioms (also called "postulates"). One of them is, for instance, that two distinct points determine a unique line.

We have just discovered that the commutative law of multiplication is not an axiom, but a theorem. It can be proved. Arithmetical axioms can be chosen in many ways. One proposition (say, A) can be chosen as an axiom and then used to prove another proposition (say, B), or proposition B can be chosen as an axiom and then used to prove proposition A. An example of a proposition that is usually assumed to be an axiom is that adding 0 to a number does not change it.

Here is a small riddle: Is the proposition $1 + 1 = 2$ an axiom or a theorem? The answer is that it is neither. It is a definition. This is the definition of the number 2.

Another Way to Learn the Commutative Law: Counting Feet

In a first-grade lesson, I demonstrated the commutative law to the children in a different way. I asked the students what was the total number of their feet.

They counted, adding another child's feet each time: 2, 4, 6, ... That day there were 23 children in class, and therefore, they calculated 23 times 2: They added $2 + 2 + 2 \ldots$ 23 times. Now I asked them how many right feet were there. 23, they said. And how many left feet? Also 23. And how many altogether? $23 + 23$, that is, 2 times 23. Since these are the same feet, the conclusion was that 23 times 2 is equal to 2 times 23.

The Rule of Change in Multiplication

A pirate found 3 treasure chests, each containing 4 coins. Thus, he had $3 \times 4 = 12$ coins. What will happen if we multiply the number of coins in each chest by 5? That is, if he has 3 chests, each containing 20 coins? The number of coins in his possession will increase 5 times, of course. He will have 60 coins, which is 5 times more than the 12 he had before: $5 \times 12 = 60$.

And what will happen if we multiply the number of chests by 5, that is, replace the multiplier, which was 3, by 15? Then, too, the result will be 5 times greater. The rule of change in multiplication is:

When the multiplier or the multiplicand is increased times a certain number, the result is increased *times* the same number.

It is very similar to the rule of change in addition, with the difference being that in multiplication the increase is "times" instead of "by."

The official name of this rule is "the associative law," since it indicates that in multiplication the grouping of the terms does not matter: $3 \times (4 \times 5) = (3 \times 4) \times 5$. On the left, we took 3×4 and multiplied the multiplicand by 5; on the right, we multiplied the entire product, 3×4, by 5 and the rule is that they are both the same.

The same rule can also be phrased the other way around: When the multiplier or the multiplicand is decreased a certain number of times, the result is also decreased that number of times.

The Distributive Law

Three brothers received 5 cookies and 2 toys each. What did they have altogether?

The answer: 3 times (5 cookies + 2 toys), which is 3 times 5 cookies plus 3 times 2 toys. That is, 15 cookies and 6 toys. This is easy to see if we express the multiplication as addition:

$$3 \times (5 + 2) = (5 + 2) + (5 + 2) + (5 + 2)$$
$$= (5 + 5 + 5) + (2 + 2 + 2) = 3 \times 5 + 3 \times 2$$

To summarize:

$$3 \times (5 + 2) = 3 \times 5 + 3 \times 2$$

This rule is called "the distributive law." It is very useful when calculating multiplication. How is 52 multiplied by 3? Since $52 = 50 + 2$, according to the distributive law the multiplication can be performed by multiplying 50 by 3 (which is 150), and adding 2 times 3 (which is 6), to get 156. Exactly as in the example of the cookies and toys!

Distribution Means Switching the Order between Grouping and Multiplication

Let's look at the following example of distribution:

$$3 \times (\diamond + \mathcal{D}) = 3 \times \diamond + 3 \times \mathcal{D}.$$

What happened here? On the left, we first grouped the sun and the moon into a set, and then repeated it 3 times. On the right, we took 3 suns and 3 moons and only then grouped them.

The fact that the result is the same means that it is permissible to first group and then multiply, or first multiply and then group. There is no difference in the result. This is a type of commutativity, and is indeed related to the commutative law of multiplication. The following, for instance, is a proof of the equality $3 \times 2 = 2 \times 3$ using the distributive law:

$$3 \times 2 = 3 \times (1 + 1) = 3 \times 1 + 3 \times 1 = 3 + 3 = 2 \times 3.$$

Distribution of Both Multiplier and Multiplicand

How do you open the parentheses in $(23 + 4) \times (5 + 67)$? The rule is that each addend of the multiplier must be multiplied by each addend of the multiplicand. The result is, therefore:

$$23 \times 5 + 23 \times 67 + 4 \times 5 + 4 \times 67.$$

One way to demonstrate this is by opening the parentheses in stages. Think of $(5 + 67)$ as a box, that is, as one number. According to the distributive law, when multiplying this number by $23 + 4$, it should be multiplied by both 23 and 4. In other words, the result is: $23 \times (5+67) + 4 \times (5+67)$. Now we use the distributive law again, for each of the two products. What we get is the result written above.

The Two Meanings of Division

The Most Interesting Operation

Division is at least as natural an operation as multiplication. It is part of everyday life — a mother divides 6 candies between 2 children, 4 children divide a deck of 52 cards. But division is more complicated than the other operations. It is also more interesting. One reason, of which many are unaware, is that it has two different meanings: *sharing* and *containment*.

In both types of division, a given set is divided into equally sized sets. But the questions asked in each type are different. In sharing division, the question is how many does each set contain; in containment division, the question is how many sets are there.

Sharing Division:
$6 \div 2 = 3$ Corresponds to $3 + 3 = 6$.

Sharing division is the type of division we are more used to in everyday life. We often divide objects equally among a known number of people. Six candies are equally divided between 2 siblings — how many candies does each one receive? The appropriate calculation is, of course, $6 \div 2$. In sharing division, $6 \div 2$ means that 6 objects are divided into 2 equally sized sets and the question is how many objects does each set contain. The result, 3, means that the two sets, each containing 3 objects, contain 6 objects altogether. In other words, 2 times 3 is 6. In the language of addition: $3 + 3 = 6$.

The names of the components of the division exercise are derived from this type of division. In the exercise $6 \div 2 = 3$ the 6 is called the dividend (since it is divided), the 2 is called the divisor, and the result, 3, is called the quotient (because 3 answers the question: How big is the quota each one received).

Containment Division:
$6 \div 2 = 3$ Corresponds to $2 + 2 + 2 = 6$.

In the second type of division, the roles of the number of sets and the number of items in each set are reversed. This time the number of items in each set is a given, and the question is how many sets are there. For instance: A mother divides 6 candies between her children, and each gets 2. How many children does she have?

Once again, the exercise is $6 \div 2$, only this time it answers a different question. In sharing division, the question was: 6 objects were divided into 2 sets, how many objects will each set contain? Now the question is: 6 objects were divided so that each set contains 2. How many sets are there?

The same can also be expressed by the question: How many times does 2 go into 6? Or how many times does 6 contain 2? The answer, 3, means that 3 sets of 2 are 6 altogether. In other words, 3 times 2 is 6, and in the language of addition: $2 + 2 + 2 = 6$. Thus, the difference between the two meanings originates in the difference between 2 times 3 and 3 times 2. In sharing division, $6 \div 2 = 3$ because 2 times 3 is 6 (or, in a formula, $2 \times 3 = 6$). In containment division, $6 \div 2 = 3$ because $3 \times 2 = 6$.

Summary

Division has two meanings.
$6 \div 2$ can be interpreted as:

(i) If 6 apples are divided into 2 equally sized sets, how many apples will each set contain? This is *sharing division*.

(ii) How many times is 2 contained in 6? In other words, if a set of 6 apples is divided into sets containing 2 apples each, how many sets will there be? This is *containment division*.

The child will encounter both meanings and therefore the distinction must be explained.

Why is Containment Division Necessary?

Sharing is the more common of the two meanings of division, since we are used to sharing with others. But the meaning of containment division is no less important. The main reason is that this is how the division operation is *calculated*. To calculate $56 \div 7$, we leap by 7s: 7, 14, 21, 28, 35, 42, 49, 56. We took 8 leaps, therefore 7 is contained 8 times in 56. This is containment division.

Those who are used to dealing cards may raise an eyebrow. When dealing a deck of 52 cards to 4 players, don't we calculate $52 \div 4$ by sharing division? After all, we are sharing the deck among 4 people!

Careful examination will reveal that there, too, the calculation is performed by containment division. The deck of cards is dealt in "rounds" — 4 cards for each round. The number of cards each player receives (the meaning of $52 \div 4$ in sharing division) is identical to the number of rounds. But the number of rounds is the same as the number of times 4 is contained in 52, which brings us back to calculating by containment division.

Containment division is also required when dividing fractions. Take, for example, the exercise $3 \div \frac{1}{2}$. Can you invent a matching arithmetical story? In sharing division, the meaning is: "I had three apples. I divided them between half a child. How many did each child receive?" "Half a child" is a confusing concept. You can explain that if half a child received 3 apples, then a whole child would receive twice as many: 6. But this explanation is difficult for children. In containment division, the question is much simpler to grasp: "I divided 3 apples between children. Each child received half an apple. How many children were there?" Or, "How many times is $\frac{1}{2}$ contained in 3?"

How many $\dfrac{1}{2}$-apples are contained in 3 apples?

Containment Division Using Leaps

Containment division can also be taught with the aid of the Jumper family, from the chapter on multiplication. Remember that the older child's step is the length of 2 tiles. How many steps will he need in order to pass 10 tiles? The meaning of this question is — how many times is 2 contained in 10? The answer, 5, is a result of containment division — dividing 10 by 2. $10 \div 2$.

Almost every class has a few students who can handle more advanced problems every now and then. The stronger students also deserve special treatment. In the second grade, I used the story of the Jumper family as such an opportunity, to introduce division by fractions. I added a small baby, who advances $\frac{1}{2}$ a tile with each step. How many steps must he take to pass 10 tiles? The appropriate exercise is $10 \div \frac{1}{2}$, expressing the number of times $\frac{1}{2}$ is contained in 10. Quite a few children knew that one tile contains 2 steps and therefore, 10 tiles contain 10 times 2, meaning 20 steps.

Division

The Rules of Change in Division

Every pirate knows that the larger the treasure, the larger the share is for each partner. On the other hand, the more partners there are, the smaller the share. A pirate is well-advised to find a large treasure and share it with a small number of partners! Here is a quantitative phrasing of the rules:

When the dividend is increased a number of times, the quotient increases the same number of times.

Reminder: In the exercise $12 \div 3$, the 12 is the "dividend" and the 3 is the "divisor." Let's assume, for example, that 2 pirates found 12 gold coins. Each will receive $12 \div 2 = 6$ coins. If the treasure is 3 times as big (so that they now have 36 coins, instead of 12), then each pirate's quotient will be 3 times bigger, that is each will receive $36 \div 2 = 18$.

If, on the other hand, the number of partners increases from 2 to 6 (times 3), each partner's share will then decrease to $12 \div 6 = 2$, three times less than before. In general:

When the divisor is <u>increased</u> a certain number of times, the quotient is <u>decreased</u> the same number of times.

Notice that we encountered a similar rule in subtraction: $12 - 2 - 3 = 12 - (2 + 3)$. Subtracting two numbers, one after the other, is like subtracting their sum. In division, dividing by two numbers, one after the other, is the same as dividing by their product!

What happens when the dividend is <u>decreased</u>, say, 3 times? Any pirate will tell you immediately that if the total treasure is 3 times smaller, each partner's share is 3 times smaller. Namely, when the dividend decreases a certain number of times, the quotient decreases that number of times. And finally, when the divisor decreases a certain number of times, the quotient grows that number of times.

Division with a Remainder

The children quickly discover that the result of division is not always a whole. Sets cannot always be divided neatly into whole parts. Seven objects cannot be divided equally between 3 children. Since the children are aware of this, it should be acknowledged right from the start. Leaving blank areas evokes a sense of uncertainty. Therefore, I believe in dealing right from the start with numbers that don't fully divide. In my experience, this poses no problem. So, what happens when you try to divide 7 objects between 3 children? Each child will receive 2 objects. One object is left over. What should be done with it? There are two options. The remaining object can also be divided between the three children, so that each child receives $\frac{1}{3}$. Or, it can be left undivided, as a "remainder." This is written as $7 \div 3 = 2R1$ (or, in European notation, 2 (1)). In words: The result is 2 with a remainder of 1.

When is each option used? This is a matter of choice, but it also depends on whether the objects can be divided into fractions. Popsicle sticks can be broken; marbles cannot. The children enjoy sorting objects according to this characteristic.

Remainders as an Opportunity for Introducing Fractions

Situations in which the remainder can be divided, like sandwiches, or pieces of paper, are a good opportunity to introduce fractions together with division. Since fractions are so close to division, it is wise to introduce them right from the start of the teaching of division. The notation for fractions such as $\frac{1}{2}$, $\frac{1}{3}$, or $\frac{1}{4}$ can be introduced already in Grade 1. When an object is divided into two equal parts it should be immediately taught that each part is a "half," and when an object is divided into 3 equal parts each part is a "third." We shall return to this point in the chapter on fractions.

First Lesson on Sharing Division

Children particularly like division skits. One should begin with sharing division. One child is invited to the front of the classroom to play the "father" (or "mother," if it's a girl). He must divide, for example, 12 straws (we call them "lollipops") between 3 children. He should be encouraged to do so in the same way cards are dealt: giving one straw to each child at a time.

Now the arithmetical story is told. The children say: A father had 12 lollipops. He divided them between his three children. How many did each child receive? We now tell the children that since the father *divided* the lollipops the operation is called "division." It is marked by a " $\div$ " sign. The operation was "12 divided by 3," and the exercise is written: $12 \div 3 = 4$. The denomination should not be omitted in the answer: 4 *lollipops*.

This should be repeated again and again, with various numbers. Then, the class is divided into groups and each group receives several objects, which they divide equally, write the exercise and report what they did to the rest of the class.

The next stage is drawing: The children draw objects (say, circles or hearts) in their notebooks or on their personal boards and divide them into equally sized groups.

In a second-grade lesson that I taught jointly with the regular teacher, I invited 6 children to the front of the classroom and asked them to demonstrate $6 \div 2$. After a brief consultation, they divided into two groups. One group came to me and the other stood next to the teacher. Because of shyness, only two children came to me, and four stuck to the teacher. This was an opportunity to phrase the principle of division explicitly. We asked the class if this was fair, and if it was really $6 \div 2$. They arrived at the wording that "division means dividing into equal parts." A negative example usually forces us to form an accurate definition of a concept.

We then asked the same 6 children to demonstrate $6 \div 3$. Having done so, they then discovered on their own that they can also divide into one group and into six groups, corresponding to the exercises $6 \div 1$ and $6 \div 6$. I asked them to demonstrate $6 \div 4$. We divided the blackboard into 4 parts, and 4 of the children dispersed, with one child standing next to each part of the board. Two girls were left. They themselves offered to bend over, so that each one would stand half next to one part of the board and half next to another. This is how we discovered that $6 \div 4$ is one and a half.

Division is the Inverse of Multiplication

After dividing 10 objects between 5 children, with each child receiving 2 objects, the inverse operation can be performed. The objects are gathered again, each child returning the 2 objects he received. The children then discover that the 10 original objects are 5 times 2. Both exercises should be written side by side: $10 \div 2 = 5$, $5 \times 2 = 10$. Then the reverse process can be done. Give 4 pencils to each of 3 children, and ask them to gather their pencils together, to form a group of 12 pencils. Write the corresponding sentence: $3 \times 4 = 12$. Then ask the children to divide the 12 pencils among them, and write side by side with the previous sentence: $12 \div 3 = 4$.

First Lesson on Containment Division

Having learned sharing division over a few lessons, it is time to get to the other type of division. I tell the kids that there is still another sort of division story. Yesterday I met a father who told me a different kind of story. He had 12 lollipops (I invite the "father" to the front of the classroom and give him 12 straws). He divided them between his children and each child received 3 lollipops. There was one thing he wouldn't tell me. Can you guess what it was? The children easily find the missing detail: How many children did this father have?

They have no problem with the answer either — the father had 4 children. But I insist we solve this systematically, without guessing. I invite one child after the other, and each one receives 3 straws. The straws run out after they are divided between 4 children.

Now is the time to ask what is the difference between the two stories. The children realize: In the first story, we asked how many each child received. In other words, we divided into 3 equal parts and asked how many *there are in each part*. In the second story, we asked how many children were there. In other words, how many *parts*. In the second story the exercise is also $12 \div 3$, yet the result is not 4 lollipops, but 4 children. It answers the question how many children there were.

First- and second-graders who were previously exposed to subtleties of meaning (mainly in subtraction) do not find this distinction daunting. But it is advisable to discern first between stories that ask about the size of the parts and stories that ask about the *number* of parts before using the terms "sharing division" and "containment division."

A Second-Grade Lesson on Dividing with a Remainder

In a second-grade class, I asked a child to divide 6 popsicle sticks between 4 children in her group. After she gave each child one stick, she was left with two sticks. We discussed with the class what she should do with them. The children offered to break each one into two and divide the 4 halves between the 4 children. This was an opportunity to teach the writing of the "half" fraction: We wrote $6 \div 4 = 1\frac{1}{2}$.

We then exchanged the popsicle sticks with coins. I asked the same child to divide 6 dimes between her 4 group mates. This time the 2 remaining coins could not be broken in half. The children offered to leave them aside. I wanted to tell the children the name for what remained — "remainder" — but the teacher stopped me. The children will find the name themselves, she claimed. Indeed, this was almost the case. After some discussion they offered the name "remnants." We told them that the official name is "remainder" and that the notation for it is $6 \div 4 = 1$ (2).

(The notation in the US is a bit different: $6 \div 4 = 1R2$. One should keep in mind that 1R2 does not denote a number, but the result of an operation, and that appreciation of its meaning demands remembering that we divided by 4.)

At this point, one of the children surprised us. Each of the 2 coins of the remainder, he said, can be changed for 2 nickels! This way each child will receive one dime and a nickel, equaling 15 pennies. We drew this option on the blackboard (we didn't have nickels) and discussed the relationship between changing money and fractions: The value of 5 pennies is half the value of a dime. We wrote the exercise in pennies: $60 \div 4 = 15$. We examined whether the result was correct. 4 times 15 is indeed 60.

Nearing the end of the lesson, we turned to personal experimentation. We paired the children and gave each pair a different number of popsicle sticks. We asked the children to divide the sticks among themselves, leaving them the choice of either breaking the extra stick (if the number did not divide equally) or leaving it as a remainder. Each child was to write the appropriate exercise. Then each pair told the class what they had done. The children were so obviously pleased (as were the teacher and I) that we ended with the children telling us what they had enjoyed about the lesson. They said that they enjoyed learning that they could either break or leave the remainder, and learning the notation for the remainder.

This was one of those enchanted lessons that make being a teacher worthwhile.

Meaning and Word Problems

Some textbooks include "word problems" as a separate chapter. The message is that this is one of the arithmetical fields or, even worse, an application. It is as if a foreign language textbook would include, amongst its many chapters, a chapter titled *How to use the studied language to describe the world*. Or an entire year would be dedicated to learning words, with their meaning taught only at the end of the year. Word problems are not one more chapter amongst the chapters of arithmetic, but the essence. They are the starting point and the end point, since they express the meaning of the operations. The meaning of arithmetic, namely how to translate real life situations into arithmetical exercises, should be taught right from the start. For those who were taught in this way, and who practiced the reverse direction as well, that is, invented arithmetical stories on their own, word problems will not be the intimidating monster that they are for most students of today.

Still, calculation is also important. After the transition from real-life situations to arithmetical exercises is complete, the result of the exercise must be calculated. This will be our next topic.

B. Calculation

What is "calculation"? The natural answer is "figuring out the result of exercises, like 234×56." But this misses the main point: Calculation means figuring out the *decimal representation* of the result. Calculations are algorithms in the decimal system, and knowing how to calculate means mastering this system. This is why students should know how to calculate, and not rely on the calculator.

There is no unique algorithm for calculating an operation; there are many ways of calculating sums and products. However, the methods currently taught in school are the result of generations of thought, and much wisdom has been invested in them. Most are based on writing the exercises vertically, so that the ones digits are one above the other, the tens digits are one above the other, and so forth.

At the base of calculations stand the addition and multiplication tables, the sums and products of numbers smaller than 10. There is much debate whether these should be memorized. The answer is a definite "yes". When you learn how to calculate 234×56 you do not want to get stuck with what is 4×6. This should be automatic, in order to be able to concentrate on the principles of the algorithm. On top of these tables, the students should be familiar with the rules that govern the operations, like the distributive law.

Another classical debate is whether "long division" should be taught in elementary school. My opinion is that this algorithm embeds such fundamental ideas, in particular that of the remainder, that it should not be passed over.

The Calculation of Addition

Continued Counting

Calculation, as mentioned, means figuring out the decimal representation of the result. To this, there is one exception, which is an elementary calculation method, appearing in addition. How do we calculate $6 + 2$? We know that $6+2$ means joining a group of 6 elements and a group of 2 elements. However, a small trick can save us a lot of effort: After all, we have already counted 6 of the elements! There is no need to repeat that which has already been done. We can start from 6, and count 2 more: 7, 8. This method is called "counting on." We shall prefer the rather less colloquial "continued counting."

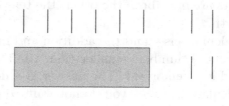

... 6, 7, 8

Understanding the principle of continued counting is considered a milestone in the child's development of thought. With good reason: It requires the ability to keep in mind a quantity, even if only for a brief moment. As fundamental as this principle is, it is not innate. It can be taught. Here is a small trick I use in the first grade to teach this principle. I draw 5 lines on the blackboard, and 3 more alongside them, and ask the children to tell me how many there are altogether. They count all the lines: $1, 2, 3, \ldots, 8$. This is fine, of course, but wasteful. Then I draw (for instance) 6 lines, and 2 more alongside them. After they count all the lines in the group of 6, I cover those with my hand and leave only the group of 2 unconcealed. Can you tell how many lines there are altogether, even though some are hidden? Now they have no choice, they can't count. But they know: 6 are hidden. If so, they can continue counting on from them: 7, 8.

Summary: Continued Counting

To calculate $6 + 2$ we start with the 6 ("this has already been counted") and continue counting 2 more, that is: 7, 8. When using continued counting it is easier to calculate $6 + 2$ than $2 + 6$.

The Addition Table

"There's a South Pole," said Christopher Robin, "and I expect there's an East Pole and a West Pole, though people don't like talking about them."

A.A. Milne, ***Winnie the Pooh***

The multiplication table is one of the main characters in arithmetic, as everyone knows. Few talk about the addition table, but it is no less important. The "multiplication table" consists of all multiples of numbers smaller than 10. The "addition table" includes their sums: $1 + 1$, $8 + 3$, $9 + 7$. They are sometimes referred to as "addition facts."

Why are the two tables so important? Because they are the basis for decimal system calculations. They are the basic tools in calculations with large numbers. They should both be memorized. Children should know the addition table in the first grade, and the full multiplication table by the end of the second grade, or in the third grade.

Summary

To add, one must know the addition table (the sum of any two single-digit numbers) and the principle of grouping tens.

To subtract, one must know the addition table, which is used to perform subtraction operations such as $13 - 5$, and also how to break tens into ones, or hundreds into tens.

To multiply, one must know the multiplication table, the principle of grouping tens and the distributive law. Division uses the same principles as multiplication, but it is slightly more difficult.

Adding Tens to Ones

When adding tens to ones, you don't actually do anything. 3 tens plus 2 ones are simply 3 tens and 2 ones. In other words: 30 plus 2 equals 32.

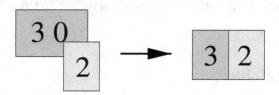

A well-known joke helps clarify the matter: *What was the color of Napoleon's white horse?* Another: *What stands on a shelf and ticks, and if it falls you need to buy a new clock?* The children discover the principle immediately: The answer lies within the question. The same is true of 30 plus 2: It is simply 32. The answer lies within the question.

Crossing the Ten 'Border'

In the exercise 8 + 5, both addends are smaller than 10, but the result is larger than 10. Out of recognition of the special importance of this type of exercise, we shall give it a special name: "crossing the ten 'border'."

Crossing the ten 'border' is important both mathematically and educationally. Mathematically, it is the basis for calculating addition and subtraction in numbers beyond 10. From the educational point of view, it is a "first time" case. Thus far the child has learned how to group tens, just for the purpose of organizing the number. This is the first time that the grouping of tens appears as a stage in a calculation. And like any first time the child does something, there is cause for applause.

First Lesson on Crossing the Ten 'Border'

Teacher (drawing a row of 14 hearts on the board): How many hearts do we have here?
Students: 14.
Teacher (writing the number 14 on the board): What does the number 1 represent?
Students: One ten.

Teacher: And the 4?

Students: 4 ones.

Teacher: That's right. 14 is one ten and 4 ones. Can you show me that in the drawing? Show me the 10.

A student comes up to the blackboard and encircles 10 of the 14 hearts.

Teacher: Very good. Inside the circle we have 10 hearts, and you can see that it is one 10. Outside the circle remain 4 hearts. Now let's make a different drawing. (*Draws 9 hearts in a row and beside them, but in a separate group, 5 hearts*). What mathematical sentence is hidden here?

Class: 9 + 5.

Teacher: That's right. Let's find out how much is 9 + 5. We have already seen that we should collect a 10. Who would like to collect a ten here? Begin with the 9 and complete a 10.

A student comes up to the blackboard and encircles 9 hearts, and adds one heart from the group of 5.

Teacher: That's correct. How many hearts remain outside of the circle?

Students: 4.

Teacher: That's right. The 5 gave 1 to the 9 to complete a 10. It is left with 4. How many are there altogether?

Students: 10 plus 4, which is 14.

This process should be repeated with arithmetical play-acting as well. Invite two children to the front of the classroom and give one 9 straws and the other 5. Remind the children that we like to collect tens. Can one child pass some straws to the other so as to complete a ten? For which child is the task of completing a ten easier? The one who has nine straws, or the one who has five straws? The children will reach the conclusion that the second child should give one straw to the first, so as to complete the first child's straws to equal 10. Having done that, he is left with 4. How many do they have altogether? Repeat this over and over, with various number pairs.

Surprisingly, this seemingly simple process requires quite a lot of previous knowledge. In fact, it requires all the knowledge learned in the first grade up to that point. First of all, it requires knowing the meaning of addition: Joining 9 hearts and 5 hearts matches the exercise 9 + 5. Then, one must be familiar with the possibility of breaking numbers: breaking the 5 into 4 + 1. It requires knowing how to break 10 into 9 + 1. There was an equation to solve: How many must be added to 9 to form a 10? And of course, it requires knowing the principle of grouping tens. Finally, it requires being familiar with the order of numbers. In the process we described, the larger of the two addends should be completed to 10: In 9 + 5, for example, the 9 should be completed to a 10, that is, 1 should be transferred to it from the 5, and not 5 transferred from the 9 to the 5. Therefore, the terms "small" and "large" must also be known.

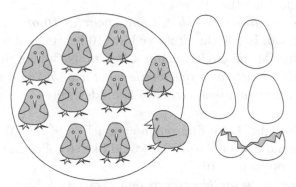

Crossing the Border of Ten: 9+5=14

From afar, crossing the border of ten seems like a simple process. Only when examined up close does its complexity, and the number of principles it contains, become obvious. If elementary mathematics seems simple, it is only because its principles are so fundamental that they are hidden from sight.

This is also a good example of the importance of systematization and knowing the preliminary stages of each subject. It can also teach us how much can be expected from first-grade students: Look how many structures of thought the children acquired over such a short period of time!

Why Does the Calculation of Addition Begin on the Right? Or: Ones are Superior to Tens

We now arrive at the calculation of addition in general. Take a look at the exercise: $92138 + 78964$. Can you tell, at a glance, what will be the thousands digit of the result? I doubt it.

But can you tell at a glance what will be the ones digit? Yes, 2. Why? Because $8 + 4$, the sum of the ones digits of both addends, equals 12, in which there are 2 ones. **All other digits of the addends represent tens, hundreds, thousands, etc.; they do not include ungrouped ones.** If so, the 2 ones will remain as they are throughout the entire algorithm. The number "2" can be written as the ones digit of the result, without fear of having to retrace our steps.

This is why calculations of addition, subtraction and multiplication begin on the right, with the ones digit. Let's demonstrate this rule with the addition of two-digit numbers:

$$26$$
$$+\,\underline{39}$$

As mentioned in the chapter *Meaning and Calculation*, the purpose of writing addition vertically is to place the ones above the ones, the tens above the tens. Again, begin on the right, with the ones digit. The sum of 6 and 9, the ones digits of the addends, is 15. Now is the time to group: 10 of the 15 are grouped into a ten, 5 are left as ones and written in the appropriate place. Notice that the ones digit will not change from now on! When we continue, we will encounter only whole tens, that will not change the ones digit.

Therefore, 5 of the 15 remained as ones. The ten grouped before is transferred to the tens (where it belongs!) How many tens are there now? 2 from the 26, 3 from the 39, and 1 more that was transferred. Altogether, 6.

The result is thus: 5 ones and 6 tens. Altogether, 65.

What would have happened had we started on the left? There were 2 plus 3 tens, equaling 5. But if the 5 had been written as the tens digit, it would have had to be changed later on, when we discovered that there was another ten to add from the addition of the ones.

This was how the ancient Chinese worked. They would start on the left and later on change what they wrote. We do the same when performing mental calculations. It makes sense: The digits on the left are more important, therefore it is reasonable to begin with them! But when dealing with large numbers, there may be too many changes to remember.

Summary

The algorithms currently used to calculate addition, subtraction and multiplication all begin on the right side of the number. The advantage of this method is that it avoids retracing steps. Once the ones digit of the result is found, it doesn't change again. Then the tens digit can be calculated without worrying about the need to take it back, and so on.

Subtraction: Loan or Reorganize?

Two Ways to Cross the Border of Ten in Subtraction

In an exercise such as $12 - 5$ the border of ten is crossed on the way down. This is the opposite of what we did in $7 + 5$. Just as crossing the border of ten on the way up is the foundation of the vertical addition algorithm, so crossing the border of ten on the way down is the foundation of subtraction.

At some point, the children should be familiar with the subtraction facts, namely all the differences that require crossing the border of ten. These facts should be memorized and then be used as part of the subtraction algorithm. However, before memorization, the way of their calculation should be taught. Here, I encountered another of the many surprises that awaited me in elementary school: I believed there was only one way to calculate differences like $12 - 5$, and discovered that there was another popular way of doing so.

The method I was familiar with included breaking the 5 into $2 + 3$, subtracting first the 2 from the 12 and then the 3. Subtracting the 2 is simple (the method is based on this fact): it brings us to 10. Then 3 must be subtracted from 10, an exercise we should already be familiar with, resulting in 7.

The second method involves breaking the 12 instead of the 5. This is simple: 12 is $10 + 2$. Now 5 should be subtracted from this sum, and we can do so by subtracting it from the 10: $10 - 5 = 5$. Now the 2 we ignored for a moment should be added: $5 + 2 = 7$.

The advantage of the first method is that it is the opposite of the addition process with which the child is already familiar. Its disadvantage is that it requires knowing how to break the 5. The advantage of the second method is that the breakdown is obvious — breaking 12 into $10 + 2$ doesn't require any effort. Its disadvantage is that it "goes down in order to go up" — subtracting and then adding, which isn't entirely natural to the subtraction process.

Which method should be used? Both should be taught. In my experience, some students are comfortable with the first method, while others do better with the second. Each child will choose the method most convenient for him. Eventually the calculation of such results as $12-5$ should become automatic, that is, memorized.

The Story of Liping Ma

Now we begin vertical subtraction, and this is an opportunity to tell a story from the recent history of mathematical education.

Since 1964, the International Education Association (IEA) has been conducting comparative mathematical tests in elementary schools in various countries around the world. Aside from providing each country with an indication of the effectiveness of its education system, these tests are also an opportunity to compare various teaching methods. It turns out that the Chinese students are doing much better than the American students in these tests. This poses a perplexing puzzle. The Americans invest more in education, their classes are smaller and, in particular, their teachers are more educated. The average American teacher spends four years at university, and so has a total of 16 years of education. The average Chinese teacher has 11 years of education: 9 in school and two in the teacher's seminar.

Liping Ma, a Chinese teacher who completed her Ph.D. in the United States, researched this issue. Her story is fascinating in itself. During the Cultural Revolution, at the end of the 1960s, as the daughter of educated parents she was sent to a remote village. The head of the village discovered her teaching skills and entrusted her with the first and second grades (as one group). When she returned to her native city Shanghai she became interested in education. She traveled to the United States and obtained a Ph.D. in education at Stanford. Her thesis compared American and Chinese teachers, in particular, with respect to their mathematical knowledge. Ma presented each teacher she interviewed with a few topics and asked how he or she would teach them. Her conclusion was surprising: The main reason for the success of the Chinese is that their teachers know elementary school mathematics better than the American teachers.

The first surprise was that there was anything to know. Ma appropriately named her book *Knowing and Teaching Elementary Mathematics*, implying that there is a lot to know in elementary mathematics, and there is a close relationship between this knowledge and the quality of teaching. The second surprise was that the advanced subjects the American teachers study at university aren't really relevant to their teaching. Knowing elementary school mathematics in depth, and respecting its complexity, is much more important since the respect is unconsciously conveyed to the students.

Vertical Subtraction: Regrouping Numbers

One of the questions Ma asked the teachers was how they would teach the exercise:

$$\begin{array}{r} 53 \\ -\ 26 \\ \hline \end{array}$$

The problem here is that 6 is greater than 3, and therefore the ones cannot be dealt with separately — only those familiar with negative numbers can calculate $3 - 6$. How can this problem be solved? The solution is commonly referred to as a "loan." The 3 "borrows" one ten from the 5 tens represented by the digit 5. There is a reason for this metaphor (which we will understand later on), but its contribution to the teaching of arithmetic is dubious. Most American teachers used this terminology. The Chinese claimed, quite justly, that it was misleading. There is no real "loan" here — the 3 doesn't really mean to return the ten it took. One Chinese teacher claimed that a loan is ordinarily open to negotiation. What if the 5 refuses to loan a ten to the 3?

The Chinese teachers used the term "regrouping," meaning that the 53 is rearranged: One ten is removed from the 5 tens and added to the 3. The result is $40 + 13$. It is crucial to understand that the 13 means there are now 13 ungrouped ones, from which the 6 ones can be subtracted, leaving 7 ones. This is written as the ones digit of the result. In the tens place there are 4 tens left, from which we subtract 2 tens, yielding 2 tens. The final result is 27.

One of the Chinese teachers related the problem to what had happened in addition (*Knowing and Teaching Elementary Mathematics, Page 8*):

> ... *They will discover the problem, that 53 doesn't have enough ones to subtract 6 from. So I will tell them: Well, today we don't have enough ones. But sometimes we have too many ones. Remember last week when we added and grouped tens? What did we do with the surplus ones? They said we grouped them into tens. ... Now, when we don't have enough ones, we can break tens back into ones. We will break one ten from the 50, and then we will have enough ones.*

The Loan Metaphor

Although the loan metaphor isn't a particularly apt one, it isn't coincidental. It conveys some truth: As in real loans, the generosity may be temporary. When a digit passes a ten to the digit on its right, it may later receive, as if in return, a larger amount — a hundred. But, unlike in a loan, it is not returned from the same digit that received it. The ten was given to the digit on its right; the hundred is received from the digit on its left.

For instance:

$$432$$
$$- \underline{198}$$

Begin on the right. To subtract the 8 ones of the subtrahend from the 2 ones of the minuend, one ten must be transferred from the 3 tens on its left. Now there are 12 ones, and subtracting 8 leaves 4. This is the ones digit of the result.

But now there's a problem with the tens, too! Of the 3 tens in the original 432 only 2 are left. From them the 9 tens of the 198 must be subtracted, and there aren't enough tens. One of the hundreds from the 4 in 432 will have to be changed and added to the 2 tens. Now there are 12 tens, from which 9 tens can be subtracted: The tens digit of the result is 3.

$$432$$
$$- \underline{198}$$
$$\square 34$$

Of the 4 hundreds in the 432, only 3 remain. From them the one hundred in 198 must be subtracted. The result is 2 hundreds. This is the hundreds digit of the result. And so, the result is 234.

The generous 3, which gave a ten to the 2 on its right, later received 10 tens (one hundred) from the 4 hundreds on its left. Does this mean that it took a loan? Not exactly. But sort of.

The Calculation of Multiplication

The Multiplication Table

A mathematician meets an old friend, and notices that he has become rich. "How did this happen?" he asks. "Gambling," answers the friend. "One night I dreamed of 6 wagons, each with 7 horses. I immediately knew that my lucky number is 6 times 7, which is 43. I bet all my money on horse number 43, and indeed it won." "But 6 times 7 is 42!" cries the mathematician indignantly. "Really?" says the friend. "Well, you're the mathematician."

In this chapter we will study the vertical multiplication algorithm, the prevalent method of calculating multiplication. It is based on two principles: knowing the multiplication table (all products of numbers smaller than 10) and grouping tens. It should be taught in the third grade.

We will begin with the multiplication table, or "plutification table" as Pippi Longstocking, the hero of Astrid Lindgren's book, calls it. Pippi claims that one can get along very well in life without it. Is it really necessary to memorize it? Do we have to remember what is 7 times 8? Isn't it enough to know, in principle, how to calculate it? (Not that Pippi is any good at that, either.) In the following chapter, we will see that there are good reasons for preferring memorization.

The first step toward memorizing the multiplication table is repeating series. For example, to learn the multiplication table of 6, the series $0, 6, 12, 18, \ldots$ should be recited, saying aloud: "0 times 6 is 0, 1 times 6 is 12 ..." The children should note that, at each step, 6 is added. Then we go backwards, starting at 60: "10 times 6 is 60, 9 times 6 is 6 less, namely 54, 8 times 6 is 6 less, which is 48," and so on. The number 9 has a special trick. 9 times 7, for example, is 10 times 7 minus 1 times 7, or $70 - 7 = 63$. Similarly, 9×6 is $60 - 6$, equaling 54. Another way to remember the multiples of 9 is to note the following rule: In 9 times 2 the tens digit is 1, in 9 times 3 the tens digit is 2, in 9 times 4 the tens digit is 3. The rule: The tens digit is 1 less than the multiplicand. For instance, in 9 times 7 the tens digit is 1 less than 7, namely, 6. And what about the ones digit? It completes the ten's digit to a 9. For instance, in 9 times 7 the tens digit is 6 and the ones digit completes the 6 to a 9. That is, it is 3.

This rule is the basis for a fingers trick of calculation of the multiples of 9. Let us again demonstrate it with 9×7. Spread all 10 of your fingers, and bend the 7th from the left (which happens to be the pointing finger of your right hand). To the left of the bent finger there are now 6 (that is, $7 - 1$)

fingers, and to the right of it 3 (which is $9 - 6$) fingers. These two numbers are the digits of the result. Personally, I prefer the $70 - 7$ method.

Multiples of 8 have a similar trick. For example, 8×6 equals $10 \times 6 - 2 \times 6$. In other words, it is $60 - 12$, which is 48.

Multiples of 5 are easy to calculate if you remember that they mean multiplying by 10 and dividing by 2. For instance, $5 \times 6 = 10 \times 6 \div 2 = 60 \div 2 = 30$.

First- and second-graders are familiar with multiples of the numbers 2, 3 and 4. So far, we've dealt with the multiples of 1, 2, 3, 4, 5, 9. What's left? Not much. Just the products of 6, 7, 8 amongst themselves (assuming the rule mentioned above for 8s is a little difficult, and isn't used). There are only six of these. First of all, the three multiples of the numbers times themselves: $6 \times 6 = 36$, $7 \times 7 = 49$ and $8 \times 8 = 64$. Other than those, there is $6 \times 7 = 42$, $6 \times 8 = 48$ and $7 \times 8 = 56$. Of the dozens of multiplications in the multiplication table, only six are left to be learned separately! These must be memorized; there is no other choice. The last one, by the way, is easy to remember, since when beginning from the right it is 5 6 7 8, namely $56 = 7 \times 8$.

Here is another rule: Multiples of 6 by even numbers are easy, since the last digit is always the multiplicand itself. For example, $6 \times 4 = 24$, the last digit is 4. Or, $6 \times 8 = 48$; the last digit is 8.

Can you discover a rule for the first digit of even multiples of 6?

Multiplication Table Week

At the end of Grade 2, or the beginning of Grade 3, I like to have a multiplication table week. The parents get in advance a note announcing this week, and are asked to participate. On the first school day of this week, the multiples of 4 are taught (the assumption being that the multiples of 2 and of 3 are already known). This is done by going back and forth on the multiples of 4, writing them, linking the different multiples to each other: If 10 times 4 is 40, what should 9 times 4 be? If 5 times 4 is 20, what is 6 times 4?). All this should not take more than an hour. Then, at home, the parents should rehearse with the child, until the multiplication table of 4 is memorized. On the second day the multiples of 6 are studied (the multiples of 5 are easy, and should be learned before), on the third day the multiples of 7, on the fourth the multiples of 8, and on the fifth day the multiples of 9. Each day the children should realize that they have fewer and fewer new multiples to learn. 9 times 4, for example, was learned on the first day, and is not new when multiples of 9 are reached. The participation of the parents makes them involved in their children's learning process, an important goal in itself.

Tip for Parents:

Memorizing is one of the easiest things to help with. Think notes on the refrigerator, a daily exercise before dinner, games. It isn't difficult to be creative in this matter.

For instance, when a child is studying the multiplication table, chant while riding in the car, with each of the passengers taking turns: "Parent: 1 times 6 is 6," and Child: "2 times 6 is 12 ...," and then in reverse: "10 times 6 is 60, 9 times 6 is 54, ... " As always, questions in the opposite direction are a good idea: Who will be the first to find a multiplication exercise resulting in 30? And an exercise resulting in 42?

Multiplying by Whole Tens

Now it is time to learn how to multiply by 10, 100, 1000, and so on. What happens to a number when it is multiplied by 10? It is common knowledge that a 0 is added to its right. For example, $23 \times 10 = 230$.

Why? Let's examine what happens to the 23 when a 0 is added on its right. The 3 ones move one place to the left and become 3 tens, thus being multiplied by 10. The 2 tens move to the left, to the hundreds place, and become 2 hundreds. They, too, are 10 times larger. Each part of the number increases 10 times so the entire number increases 10 times.

Similarly, multiplying by 100 adds two zeros to the right of the number, multiplying by 1000 three zeros, and so on.

Vertical Multiplication

Stage One: Multiplying by a Single-Digit Number (Smaller than 10)

We now have all the tools to perform vertical multiplication. However, we will study it one stage at a time. The first stage is multiplying by a single-digit number, for example, 234×7. First of all, remember the meaning of 234, which is $200 + 30 + 4$. By the distributive law we should therefore multiply 7 by 200, 30 and 4, and add the results.

A useful observation, which has already been made, is that single ones will be obtained only from the multiplication of 7 by 4, because the multiplication of 7 by 30 and by 200 will provide only tens and hundreds. The ones digit will thus come only from 4×7, namely 28, which has 8 single ones. We now know the ones digit: It is 8.

So, we write 8 **in the ones digit place**. But the result was 28, so there are still 2 more tens to be added to the result. These must be kept in mind. It is customary to write them above the tens digit of the multiplier, above the 3 in 234.

Now, move on to the tens and multiply 30 by 7. Here we use a shortcut: The 3 (the second digit of the number 234) is multiplied by 7, while we remember that it means 3 tens and not 3 ones. The result is 21 tens. But remember there are still 2 tens from the previous round. Thus, there are now 23 tens, which are 3 tens and 2 hundreds. The 2 hundreds are kept for the next stage, since they do not affect the tens digit. Now there is no problem to write **3 in the tens digit place**. There is still 7 to multiply by 200, but the result will be in hundreds and will not alter the tens digit!

Remember, of course, that there are still 2 hundreds left from the previous round, and these pass on to the next round.

Finally, the 2 hundreds of the number 234 are multiplied by 7. The result is 14 hundreds. Add the 2 hundreds left over from the previous round, so that altogether there are 16 hundreds, which is **1 thousand and 6 hundreds**.

Collecting everything, the result is 1638.

Stage Two: Multiplying by Tens

Next, we learn how to multiply by whole tens, namely by numbers like 20. How does one calculate, say, 234×20? The answer, as in the kettle joke, is that "this has already been solved." Whoever knows how to calculate 234×2 knows also 234×20. We know that $234 \times 2 = 468$. (This is an easy

multiplication task — every digit is multiplied by 2, there is no carry over.)
But then:

$$234 \times 20 = 234 \times 2 \times 10 = 468 \times 10 = 4680.$$

We just multiplied by 2, and added a zero on the right!

Stage Three: Joining the First Two Stages

We can now join the two previous stages together. In the examples, we
calculated 234×7 and 234×20. Now let's calculate 234×27.

According to the distributive law, 234×27 is equal to $234 \times 7 + 234 \times 20$.
(234×7 appears first because that is how long multiplication is performed —
first the number is multiplied by the ones digit and only then by the tens.
This is a matter of convention and not of necessity.) The result of the first, as
recalled, is 1638. The second is 4680. To add them, we place them, as usual,
one above the other. In fact, to improve organization we write the result of
234×20 below the 1638 right when this multiplication is performed. Thus,
the entire algorithm looks as follows:

$$
\begin{array}{r}
234 \\
\times \quad 27 \\
\hline
1638 \\
+4680 \\
\hline
6318
\end{array}
$$

That's it. That's the algorithm. However, at this point there's a small
turnabout: The ones digit, 0, is erased from the number 4680. The algorithm
then takes the shape:

$$
\begin{array}{r}
234 \\
\times \quad 27 \\
\hline
1638 \\
+468 \\
\hline
6318
\end{array}
$$

Why is this allowed? We know the 0 digit acts as a placeholder, to clarify
that there are 4 thousands, 6 hundreds and 8 tens, and not 468. Writing
the exercise in columns takes care of this. The numbers are written in the
correct columns. The digit 8, for instance, appears beneath the tens digit in
1638, and therefore we know that it means 8 tens and not 8 ones.

If so, the 0 can be discarded. But this isn't really necessary. So why is it
still customary? It's because the 0 is a result of multiplying by 20. However,

the number we are multiplying by, 27, contains a 2 and not a 20. If you recall, this is exactly what we did. We multiplied by 2, and only then added the 0 on the right. Therefore, the 0 wasn't there in the first place and there is no real reason to add it. Some tend to leave the 0. This may be wise. It makes things appear less magical.

What happens when there are hundreds in the multiplicand? Take, for example, 234×527? To the result of 234×27 calculated above, 234×500 must now be added. We already know how to calculate this: Multiply 234 by 5 and add two zeros. However, we have already learned our lesson: The two zeros aren't really necessary. Their job is to push the number two places to the left. We can push it on our own, without their help!

Why This of All Methods?

234×27 can also be calculated in a way that is easier to describe:

$$234 \times 27 = (200 + 30 + 4) \times (20 + 7)$$
$$= 200 \times 20 + 200 \times 7 + 30 \times 20 + 30 \times 7 + 4 \times 20 + 4 \times 7.$$

This, as recalled, is the distributive law of both the multiplier and the multiplicand: Each of the addends of the multiplier is multiplied by each of the addends of the multiplicand. We know how to calculate each of the six products. In fact, we calculated all six using the regular algorithm as well, but organized things a bit differently there.

Some mathematics educators claim, seemingly reasonably, that it is enough to teach this method. There is no need for much practice or many multiplication exercises to understand it. It may be a bit cumbersome, but the idea isn't that the children actually use it to calculate in the end, but only understand the principle and then use a calculator. Adults use calculators for such calculations, too!

The Advantage of the Classic Algorithm

Nevertheless, it is better to rely on the wisdom of generations which is embodied in the old algorithm. It is more organized and requires less writing: In the example above, the classic algorithm requires the addition of two numbers only, instead of six! This is because some of the addition operations are performed mentally.

Because the other method is so cumbersome, using it is almost certain to mean very little practice. This is unwise. Educational revolutions have failed for neglect of practice. Principles are assimilated only through hard work.

There is another issue here. Avoiding practicing the algorithm assumes the use of the calculator. Once the child gets used to the fact that there are things beyond his capabilities, and that he can then lean on the calculator, he will learn to use it more and more. It is the experience of leaning on an external calculation aid (and in this case an aid that doesn't only assist, but does the entire job) in itself that is harmful.

Memorize or Calculate Anew?

In Praise of Memorization

"Understand, don't recite," say educators over and over. There isn't actually that much to memorize in elementary mathematics. Only twice during his studies does the child need to remember facts by heart: the addition table in Grade 1, and the multiplication table in Grade 3. There is indeed no way to bypass the memorization of these two. If a child writing an essay is still not sure how to write a "g," it will eventually hold back his writing. When a child who does not know the addition table by heart encounters a sum such as $28 + 17$, he will waste time calculating $7 + 8$ instead of learning the new principles required for this exercise. The same is true of multiplication. When calculating 24×7, the calculation of 4×7 and 2×7 should be automatic.

The question is how to acquire this knowledge by heart: make the effort to memorize, or calculate anew each time and expect the memory to form with time? My initial inclination was toward the second possibility. Calculating again and again is a good opportunity for mastering the principles of calculation.

It took two of my own children who preferred (or maybe their schools preferred) this method, and refused to memorize, to make me realize that this simply doesn't work. It is possible to calculate 8×7 many times without remembering the result by heart. I should have known this in advance, since I know that a telephone number can be dialed dozens of time without being committed to memory. If no intentional effort is made to remember the number, it just doesn't stay. It is transferred from the eye to the fingers, without leaving any trace in consciousness.

Why? Psychology has an interesting answer.

Two Types of Memory

What has four legs, a tail and goes "woof"? — A dog. — Oh, someone told you already!

"Someone told you" or "you already knew" shouldn't be treated lightly. It is a type of memory. The psychologists call it "declarative memory." It includes things you said to yourself explicitly, or others told you and you declared to yourself that they are indeed so.

The "dog" joke provides two examples of declarative memory. One is "you already knew", the other is that a dog goes "woof." Only an American dog goes "woof." Israeli dogs go "how-how" and Japanese dogs go "hip-hip." Each language made a different, pretty arbitrary, choice. In fact, dogs don't make any of those sounds. You were told that they do, and have since kept the information in your declarative memory.

Another type of memory is the memory of events: things that happened to you, or things you did. There is little connection between the two. Something that happened to you isn't always put into words and isn't always remembered. To pass from one to the other requires a declaration and a deliberate effort to learn.

There is no alternative: Some things need to be memorized. It's not that difficult. Very little practice saves a lot of time in the long run and enables one to move on to the next stage. Trying to calculate anew each time is like living in a foreign country, refusing to learn its language, and instead walking around with a dictionary your entire life.

For example, notes with multiplication table facts can be put up on the refrigerator door, or a daily exercise can be given.

Calculation of Division Begins on the Left

Long Division and the "Math Wars"

How is $1368 \div 72$ calculated? Nowadays, there is an abridged algorithm:

Take the calculator out of your pocket, punch in the appropriate numbers, and press the "=" button. In the past, it was performed differently, using a method called "long division," so named because of its many stages. Those who studied it acquired a good sense of numbers and a profound understanding of the decimal system.

Of the algorithms for calculating the four arithmetical operations, long division is the most difficult. Many educators have drawn the conclusion that it is unnecessary. They claim that most children don't understand it. Division, they claim, should be calculated using a calculator, just as we use a calculator to calculate square roots and don't do them ourselves. On the other hand, many believe that long division contains principles essential to understanding the decimal system. And so the question of teaching long division became a bone of contention, the symbol of the disagreement between advocates of the use of calculators and their opponents, and a major dividing issue in the so-called "math wars."

There is a great benefit to understanding long division. Many principles are imparted through it: the meaning of division, estimation, as well as a good understanding of the decimal system. Aside from all these, the sense of mastery is important. It is important that the child knows how to calculate so common an operation on his own. And, finally, the algorithm isn't really that complicated. In the following sections, we will try to detail its stages.

One of the achievements I am most proud of is that I was one of a group of mathematicians who persuaded the Israeli ministry of education to abandon a calculator-tilted curriculum, and implement the teaching of long division in the curriculum that replaced it.

First Lesson on Long Division

Long division is called "long" because, when written on paper, it usually turns out to be quite a long column of numbers. What is this vertical way of writing? Multiplication is written vertically so that we can vertically add the different components of the product. In division, the reason is different: Here we do vertical subtraction. The subtraction is done so as to find that part of the dividend (the number to be divided) which has not been divided so far or, in other words, the remainder.

This is the main idea of the algorithm, and it must be taught first. To teach it, it is best to start with a very simple, almost trivial example. In fact, an example which doesn't really require long division at all. Long division is required when the result is greater than 10, but the principle can be taught with results smaller than 10.

In a third grade I started the teaching of long division by returning to division with remainder. I took 7 pens, and asked a girl to divide them between 2 students. She gave each 3 pens, and kept the remaining pen as a remainder. How many pens did you really divide? 6, she said. How do you know? Well, there were 2 students, each got 3, so of the 7, I divided

$$2 \times 3 = 6 \text{ pens.}$$

Let me tell you something new, I told the class. There is another way of writing division. Instead of $7 \div 2$ people sometimes write:

$$2 \overline{)\, 7}^{\,3}$$

We write the divisor on the left, and the result on top. Now, what did we do to find out how many pens were really divided? We multiplied 2 by 3. Let's write that below the 7, as follows:

$$\begin{array}{r} 3 \\ 2\overline{)\,7} \\ 6 \end{array}$$

And what did we do to find out how many were left, that is, the remainder? We subtracted, like this:

$$\begin{array}{r} 3 \\ 2\overline{)\,7} \\ -6 \\ \hline 1 \end{array}$$

Why did we write it vertically? After some discussion we reached the conclusion — in this case, it wasn't really necessary. But if the numbers were larger, then writing them one above the other would facilitate the subtraction.

All this seems as simple as a chess game on a 2-by-2 board. Is it really necessary? Yes, of course. Principles should be broken apart, and each ingredient should be taught on its own. Even the simplest part has to be explicit.

Didactic Suggestion: Begin with Dividing by 2

The performance of long division requires understanding of the outline of the algorithm, and calculating the division operations involved in each stage. To facilitate the understanding of the former, it is wise to make the second part as easy as possible. This is best done by starting with division by 2, in which the actual division operations are easy. They only involve dividing numbers smaller than 20 by 2, which is within easy reach of most children.

First Case: Dividing by 2 When All Digits are Divisible by 2

Take a look at the exercise $68 \div 2$. We know that $8 \div 2 = 4$, and $60 \div 2 = 30$. Since $68 = 60 + 8$, according to the distributive law of division

$$68 \div 2 = 8 \div 2 + 60 \div 2 = 30 + 4 = 34.$$

What happened here? We simply divided each digit by 2! We separately divided the tens digit, 6, by 2 and the ones digit, 8, by 2. This is possible whenever all digits are divisible by 2. For example, $8642 \div 2 = 4321$.

Long Division Begins on the Left!

How is the calculation of an exercise such as $758 \div 2$, in which the dividend has a non-even digit, 5, to be performed? Here we reach the rule after which this chapter is named: Start on the left. So, the rule is opposite to the one we use in addition and multiplication. But the reason is the same: trying to avoid the retracing of steps. There is a digit in the result that can be calculated immediately, without fear of having to change it at a later stage. In division it happens to be the leftmost digit.

How many ones will there be in $758 \div 2$? Not necessarily $8 \div 2$, namely 4. (We will see why later. We will actually have to divide 18 by 2, resulting in 9.) But it is clear how many hundreds there are. There are 7 hundreds in the dividend, and when 7 hundreds are divided by 2 the answer is 3. The hundreds digit of the result is therefore 3. Nothing will change this from now on: Dividing the tens by 2 will not add hundreds to the result.

What is Done With the Undivided Hundred?

What happens now? Not all 7 hundreds were divided. Only 6 of the hundreds were divided, and 1 hundred is left undivided, a remainder. When it is divided, it won't yield hundreds, but tens. To know how many tens, the one hundred needs to be changed for 10 tens. And 10 tens can be divided by 2.

However, 758 already has 5 tens, represented by the second digit. When they are added to the 10 tens changed from the hundred, there are 15 tens altogether. Divided by 2, they are 7 tens. This will be the tens digit of the result.

Not all tens have been divided. Of the 15 tens, only 14 were divided. One ten is left undivided. 1 ten cannot be divided by 2 as a ten. It needs to be changed and added to the ones. Along with the 8 ones of the ones digit we now have 18 ones. Divided by 2, there are 9 ones, without remainder. Thus, the final result is 3 hundreds, 7 tens and 9 ones, 379.

Writing It Up

We have already had a glimpse of the customary way of organizing what we just did. Begin with the hundreds digit. 7 hundreds divided by 2 are 3 hundreds, which are written above the hundreds digit of the dividend, thus:

$$\frac{3}{2\,\overline{)\,768}}$$

Why is the divisor, 2, written to the left of the dividend, 748? There is no reason. In Israel it is written on the right. The first time I taught in the Unites States I wrote division as I had my entire life, innocently believing in the universality of writing the divisor on the right, and the entire class burst out laughing.

Let's get back to the calculation. The problem now is that not all 7 hundreds were divided, but only 6 of them. How do we know? If 3 is multiplied by 2 the result is 6, meaning that only 6 hundreds were divided. Write these under the 7 and subtract. That's how we know how many hundreds are left undivided:

$$
\begin{array}{r}
3 \\
2\,\overline{)\,758} \\
-6 \\
\hline
15
\end{array}
$$

7 − 6 equals 1, one hundred left undivided and as mentioned before, it is changed for tens. The 5 tens from the 758 are written next to it (see bottom

line), and together there are 15 tens. When divided by 2, they are 7 tens. These are written in the tens place of the result: 7 tens.

$$
\begin{array}{r}
37 \\
2\ \overline{)\ 758} \\
-6 \\
\hline
15
\end{array}
$$

How many tens were really divided? We can find out by multiplying the 7 by 2. The result is 14, and when subtracted from the 15, one ten remains. This is the ten that was left undivided.

$$
\begin{array}{r}
379 \\
2\ \overline{)\ 758} \\
-6 \\
\hline
15 \\
-14 \\
\hline
18
\end{array}
$$

To the 1, representing the ten that needs to be divided, the 8 ones are added, resulting in 18. Divided by 2 there are 9, this time with no remainder. One can check that there is no remainder by multiplying 9 by 2, and subtracting the result from 18, written as:

$$
\begin{array}{r}
379 \\
2\ \overline{)\ 758} \\
-6 \\
\hline
15 \\
-14 \\
\hline
18 \\
-18 \\
\hline
0
\end{array}
$$

After achieving full proficiency in dividing by 2, the road is clear. The children should practice long division by 3 and by 4. Later, more complex cases can be introduced. Here is another example: $632 \div 7$. This time the emphasis will be on the recipe for division and the explanations will be shorter.

How Many Hundreds are There in $623 \div 7$?

There are 6 hundreds in 623. Therefore, there are $6 \div 7$ hundreds in $623 \div 7$, which is 0 (with a remainder of 6, but this remainder, as recalled, turns into tens in the next stage).

The hundreds digit of the result is 0. It is written thus:

$$\frac{0}{7 \overline{)\ 623}}$$

The 0 isn't usually written, but it contains an educational element, a reminder of our attempt to divide hundreds. Therefore, we will leave it until the algorithm is complete.

How Many Tens are There in 623 ÷ 7?

The 6 hundreds of 623 are still undivided. In the next stage, they are changed for tens. Six hundreds are 60 tens. Along with the 2 tens of the number 623 there are 62 tens. Therefore, the number of tens in the result is $62 \div 7$, which is 8. Write 8 in the tens place.

$$\frac{08}{7 \overline{)\ 623}}$$

Multiplying 8, the number of tens in the result, by 7 will show us the number of tens that have thus far been divided: 56. This is written under the 62 tens that were supposed to be divided. The difference, $62 - 56$, is the number of tens that remain and need to be divided. It is written thus:

$$\begin{array}{r} 08 \\ 7 \overline{)\ 623} \\ -56 \\ \hline 6 \end{array}$$

All that's left is to calculate the ones digit of the result. There are 6 tens left to divide, and they cannot be divided by 7. This is not by chance; all the tens that could be divided have already been divided. The next stage is to change the tens for ones, and add the 3 ones from the 623 to them. This yields 63 ones altogether. These can be divided by 7, resulting in 9, with no remainder. This is the ones digit of the result. The result is, therefore, 89. Written as an algorithm:

$$\begin{array}{r} 089 \\ 7 \overline{)\ 623} \\ -56 \\ \hline 63 \end{array}$$

Checking that there is no remainder is done by multiplying 7 by 9, and subtracting, written as:

$$
\begin{array}{r}
089 \\
\hline
7\,)\;623 \\
-56 \\
\hline
63 \\
-63 \\
\hline
0
\end{array}
$$

C. Fractions

By "fractions" I mean here ordinary fractions, like $\frac{1}{2}$ as opposed to decimal fractions, like 0.25.

What is a fraction? I often ask this question in schools, even high schools. Students and teachers alike start stammering something like "part of a whole" (tough already the ancient Greeks knew that a part of a whole need not be a fraction — there are numbers that cannot be expressed as the quotient of two whole numbers), or "numerator and denominator" (which does not explain much). The most accurate answer I get is "the quotient of two integers (whole numbers)". This is precise, but as we shall see, it is not the definition of "fraction", and does not help understanding it.

Here is the real definition, the one that the teaching of fractions should follow:

A fraction is the combination of division and multiplication, in this order.

For example, "two thirds" of something, the "whole", is obtained by dividing the whole by 3 — this yields one third; and then multiplying the third by 2, which plainly means repeating it twice. These operations can be performed in the reverse order — first multiplying the whole by 2, and then dividing the result by 3, but this is already a theorem, that we shall later see why it is true, and not a definition. The definition is: first divide, then multiply.

So, there are two steps. In the first, a fraction of the form $\frac{1}{2}$, a $\frac{1}{3}$, a $\frac{1}{4}$ etc., namely with numerator 1, is found. This is just division: $\frac{1}{3}$ of a whole is just this something divided by 3, nothing else. Then comes the multiplication. You may well ask: why don't we combine the other pair of operations, addition and subtraction? Why, for example, don't we subtract 3 and add 2, to get some new object? Because, of course, subtracting 3 and adding 2 results plainly in subtracting 1. The result is not a new operation.

Since fractions are formed from division and multiplication, it is easy to multiply and divide them. Much easier than adding and subtracting them. So, my recommendation is to start the study of operations on fractions from these two operations, multiplication and division.

Division and Fractions

*Even now, beneath us, the world is a fraction line Don't fear, see how beyond
That same line now disappears The common denominator.*

Yehuda Amichai, "Up On the Acorn Tree," **Poems**

The Most Complex Topic in Elementary Arithmetic

In the Louvre museum in Paris there is an interesting document from the 15th century: a correspondence between a concerned father and his mathematician friend. The father asks to which university he should send his son to study. The mathematician answers that University A is OK, but if he really wants his son to understand fractions, he should send his son to University B.

Yes, only 500 years ago, fractions were taught at the university level! And indeed, fractions (and the related topic of ratio problems) are by far the most complex topic studied in elementary school. Their depth is indeed on a par with that of many university topics.

Numerator, Denominator, and the Whole

The fraction $\frac{2}{3}$ is composed of the numerator, 2 (above the fraction line), the denominator, 3 (below the fraction line), and the fraction line itself. $\frac{2}{3}$ is an example of a simple fraction, namely a fraction in which the numerator is smaller than the denominator. There are also improper fractions, where the numerator is greater than, or equal to, the denominator, such as $\frac{5}{3}$ or $\frac{3}{3}$. There are also mixed numbers, which are combinations of an integer (whole number) and a fraction, such as $2\frac{3}{5}$, namely two and three fifths.

The name of the denominator is derived from the fact that it is a denomination ("nom" means "name"): It indicates what part is taken from the whole. For example, the denominator "3" in the fraction $\frac{2}{3}$ indicates that we are dealing with thirds. The numerator, as indicated by its name, enumerates the number of such parts.

A fraction is taken from a whole, and the choice of the whole is arbitrary, on the one hand, and crucial on the other hand. Look for example at the following figure:

Is this a half or one? It all depends on the definition of the "whole."
If the whole is a cake, this is a half. If the whole is half a cake, this is one.

We tend to say that this is a "half". But this is just because our mind completes it to a whole circle (say, cake). If the "whole" is just this shape, then this is one whole. And if the whole is two complete circles ("cakes") then this is just a quarter.

And just as in counting we formed the pure number 2 from two apples, two pencils and two cats, in fractions, too, we form the pure fraction $\frac{2}{3}$ as an abstraction of taking 2 out of 3 equal parts of an apple, or of a small circle, or of a large circle, or of a group of 30 children. So, just as in numbers, fractions not only count wholes. They become numbers with independent existence.

Summary

The numerator counts (enumerates!) the number of parts. The denominator indicates the type of part that is being counted, or the number of equal parts into which we divided the whole.

First Lesson on Fractions

To establish a connection between division and fractions, I like to begin not with the division of one object, but of many. After doing divisions such as $6 \div 2$, where there is no remainder, I let 2 children divide between themselves, say, 7 popsicle sticks. Each child receives 3 sticks. We then discuss what to do with the remaining stick. The children offer to break it into 2. They also know how to call each part: "a half."

Then we introduce the notation: $\frac{1}{2}$. The children themselves guess its origin: The denominator is 2 since we divided the whole into 2 parts; the numerator is 1 because we took 1 part. They then discover on their own, much to their pride, the notations $\frac{1}{3}$ and $\frac{1}{4}$.

Fractions with numerator 1 are to be introduced first. They are called "Egyptian fractions" since the Egyptians used only such fractions. The fractions $\frac{1}{2}$, $\frac{1}{4}$, and possibly even $\frac{1}{3}$ can, and probably should, be taught at the end of first grade. In order for the children to internalize the meaning of the fraction, they must experiment with division over and over, with concrete examples and drawings: finding a half, a third and a quarter of different sets and geometrical shapes.

A third: of a small cake, of a large cake, of a set of 6 apples, of 2 rectangles.

The second stage is understanding the role of the numerator: What are two thirds or three quarters? It is customary to begin with parts of geometrical shapes: When dividing a pizza into 3 equal parts, two thirds are 2 pizza slices. But it is best to move on to parts of a number as soon as possible: A third of a group of 15 pencils is 5, and therefore two thirds are 2 times 5, that is, 10. In my experience, the children never tire of finding fractions of various wholes of all types.

Historical Note

Where did the notation for fractions originate?

It was the Egyptians who came up with the idea of the vertical structure, with the denominator on the bottom. The Egyptians only had fractions with a numerator of 1, such as $\frac{1}{2}, \frac{1}{3}, \frac{1}{4}, \ldots$ (Such fractions are therefore called "Egyptian fractions.") However, their notation was different: Instead of the numerator they drew an oval, and they didn't write the fraction line.

Fractions with general numerators were first used in India. The Indians didn't use a fraction line either, but simply wrote the numerator above the denominator.

The fraction line was introduced by the 13th century mathematician Leonardo from Pisa (also known as Fibonacci).

Lesson on Wholes

Draw half a circle (half a pizza) and ask the child to mark half of it. This will result in a quarter of a circle, of course. But how did the quarter suddenly become a half? Did the child mark a quarter, or a half?

Then draw a large pizza and a small pizza. Ask the child to draw half of the first and half of the second. The drawings aren't the same, of course. How can they both be a half?

This is a good starting point for an important discussion: *The nature of the half depends on the chosen whole.* If the whole is an entire pizza, then half of it is half a pizza. If the whole is half a pizza, then half of it is a quarter of a pizza. If the whole is 20 flowers, then half are 10 flowers. In fractions, just as in the case of numbers, the denomination makes all the difference.

When we discussed the importance of the denomination in counting, we mentioned the trick of asking a child to give two, leading to a discussion of the question: "two of what?" Do the same here. Ask the child to draw a half. He will have to think "half of what?"

Summary

The "whole" in fractions is like the denomination in counting. Just as the denomination tells us what we are counting, the whole tells us what the fraction is taken from.

In one second-grade lesson, after the children had experimented with division and some fractions, I divided them into groups and gave each group a task: divide an object into 3 equal parts. One group received a circle, the other 2 circles, the third a square, the fourth 2 popsicle sticks, and the fifth a drawing of 7 sticks.

They were extremely excited. Some didn't know how to divide a circle, but one child, whose father owned a pizzeria, did. After the lesson I moved on to another class and returned to the first for a visit an hour later. I discovered that the children had demanded to continue — they divided triangles, trapezoids and ovals into 3. Not all shapes could be easily divided into 3 equal parts, but the children insisted!

Why are the Teaching of Fractions and Division Separated?

A mistake that permeates the teaching of fractions all over the world, and in all textbooks, is its separation from the teaching of the operation of division. Fractions are usually taught much later than division, and presented as a totally new topic. Why is this so?

The reason is that division is first taught of *numbers*, like 12÷3. Fractions are almost ubiquitously first taken of *shapes*, very often, of circles (pizza slices). Very often the children reach the meaning of "a third of a set of 12 elements" only much later, and find it hard to connect it with the pizza slices.

There are reasons for this choice. The idea behind first taking fractions of shapes is that taking a fraction of a number demands abstraction. It means that the "whole" is a set: Taking a third of 12 means that the whole is a set of 12 elements. And it is easier to consider a circle, or a rectangle, as a whole, than taking a set and declaring it to be one unit.

Still, both choices are erroneous. The teaching of division should from the start include the division of shapes: Divide a rectangle, a circle, or two rectangles into three equal parts. And it should at the same time include dividing sets into three parts: a set of 3 apples or a set of 12 children. Each of the parts should be named a "third" of the whole, and two parts should be named "two thirds." This means that fractions are taken, from the very start, also of sets, that is, of numbers.

When the Whole is a Set

When the whole is a set, the calculation required is a fraction of a number. For example:

60 children went on a trip. $\frac{2}{3}$ of them were girls. How many girls went on the trip?

A $\frac{1}{3}$ of 60 is obtained by dividing the whole, which is a set of 60 members, into 3 parts, and taking one part. Since 60 ÷ 3 = 20, a $\frac{1}{3}$ of 60 is 20. $\frac{2}{3}$ of 60 is (as the name "two thirds" indicates) two such parts, namely 2 times 20, which is 40.

Similarly, $\frac{3}{5}$ of 200 is obtained by dividing 200 by 5 (resulting in 40) and multiplying by 3, which yields 3 × 40 = 120.

In general, to find a part of a number, first divide the number by the denominator and then multiply by the numerator.

Now we must discuss the reverse: If a part of a number is given, what fraction of the number is it?

There are 32 students in a class. 20 of them are boys. What part of the class do the boys constitute?

Children find this type of question harder than the previous one. They will easily answer the question: "What part of 20 is 10?" Answer: A half. But a question like the one above is difficult for them.

As in many other instances of difficulty, the reason here is an attempt to take two steps at once. This is a classic example of how breaking a problem into stages makes it easy. The first stage should be: What part of a class of 32 children does one child constitute? This is easy: One of 32 is $\frac{1}{32}$. The second stage is: "If so, what part of the class do 20 children constitute?" $\frac{20}{32}$ of the class. This can be reduced by dividing the numerator and the denominator by 4 (see next chapter): $\frac{20}{32}$ can also be written as $\frac{5}{8}$. Similarly, one day is $\frac{1}{7}$ of a week, and hence 5 days are $\frac{5}{7}$ of a week, and so forth.

Finding the Whole from Its Part

So far, we found the part from the whole. To find $\frac{2}{3}$ of a number, it is multiplied by 2 and divided by 3. In the other direction, one knows the part, and wants to find the whole. For example, if we know that $\frac{2}{3}$ of a class are 26 students, how many children are there in the class?

This may sound hard, but remember: When something is difficult, it means that we did not break it into stages! "Ask me a simpler question," as my daughter taught me. The first question should be: If $\frac{1}{3}$ of a set is so and so, how large is the set? For example, if $\frac{1}{3}$ of a class is 7 students, how many students are there in the class?

The answer to this is obvious. The whole is 3 times the third, which is 3 times 7, namely 21. And if a fifth of the price of a CD-player is 31 dollars, then the price itself is 5 times $31, namely $155.

Let us now return to the original question, of finding the number of students in a class, given that $\frac{2}{3}$ of the class is 26 students. We can break the problem into two stages. First, if $\frac{2}{3}$ of the students are 26, how many is $\frac{1}{3}$? $\frac{1}{3}$ is a half of $\frac{2}{3}$. Therefore, if $\frac{2}{3}$ of the class is 26, $\frac{1}{3}$ is $26 \div 2 = 13$. The second stage: If $\frac{1}{3}$ of the students is 13, how many students are there in the class? This we have already solved, it is 3×13, namely 39 students.

To find the whole from $\frac{2}{3}$ of it, we divided by 2 and multiplied by 3. This is the exact opposite of the operations required to find $\frac{2}{3}$ of the whole. This is not surprising, since finding the whole from its part is the inverse operation to that of finding the part from the whole.

Here is another example. It is given that $\frac{5}{7}$ of a group is 40, how big is the group? $\frac{1}{7}$ of the group is $40 \div 5$, that is 8, since $\frac{1}{7}$ is 5 times less than $\frac{5}{7}$! And if $\frac{1}{7}$ of the group is 8, then the entire group is 7 times larger, namely, $7 \times 8 = 56$.

The operations included here were: division by 5 (the numerator of $\frac{5}{7}$) and multiplication by 7 (the denominator). To find the whole from its part, multiply by the denominator and divide by the numerator.

Fractions are the Tool for Handling Division

Question: Why do Jews answer a question with a question?
Answer: And why shouldn't Jews answer a question with a question?

Why are fractions so important? Let's answer the question with a question: Why are negative numbers necessary? Everyone knows the answer to that: So that the note pulled out of the ATM can indicate an overdraft. Negative numbers enable you to pay $5000, even when there are only $2000 in your bank account, leaving you with $2000 − $5000 = −$3000. Negative numbers enable the subtraction of a larger number from a smaller one.

Similarly, fractions enable dividing a smaller number by a larger number. For instance: $2 \div 3 = \frac{2}{3}$. We invented a new type of number, called "two thirds," which is the result of the division of 2 by 3, just as we invented a new number "minus 1" so that we could write the result of the exercise $2 - 3$ as a number: -1.

The following example illustrates the fact that fractions are actually division. How do you divide 2 cakes between 3 children? First, divide the first cake between the 3 children. Each child will receive $\frac{1}{3}$. Now divide the second cake. Again, each child will receive $\frac{1}{3}$. Altogether, each child will receive $\frac{2}{3}$.

This demonstrates that $2 \div 3 = \frac{2}{3}$. In other words, the result of the division is a fraction, in which the numerator is the dividend and the denominator is the divisor. Therefore, the fraction line can be viewed as a division sign: $\frac{2}{3} = 2 \div 3$.

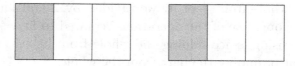

Divide each cake between 3 children. Altogether each child will receive $\frac{2}{3}$.

But why is the notation necessary? Why not just use the notation $2 \div 3$?

The reason is that $2 \div 3$ is an *operation*, like $2 + 3$, and we are looking for a number that is the *result* of an operation. Before the invention of the fraction, the result of the division of 2 by 3 could not be discussed. Ever since the invention of fractions, it can — the result is the number $\frac{2}{3}$.

Fractions Turn Division into Multiplication

In What Order Should Operations with Fractions be Taught?

In whole numbers, the teaching of addition and subtraction precedes multiplication and division. The reason is obvious: In such numbers addition and subtraction are more basic and simple. Partly for reasons of persistence, this order is usually carried over to the teaching of operations with fractions. This is not completely justified since fractions, being numbers expressing division, get along better with multiplication and division than with addition and subtraction. Clearly, calculations such as "two sevenths plus three sevenths," which are not really different from "two apples plus three apples," should be taught right at the beginning of the teaching of fractions. But when the addition and subtraction of fractions involve finding a common denominator, they are certainly harder than multiplication and division of fractions, and hence should rather be taught later. Also, the expansion and reduction of fractions, which is necessary for understanding the notion of the common denominator and finding common denominators, are closely related to multiplication and division of fractions. For this reason I adopted here the less orthodox approach of starting with multiplication and division.

Multiplying a Fraction by a Whole Number

What is 4 times $\frac{2}{3}$? The fraction $\frac{2}{3}$ means "2 thirds," and just as 4 times 2 apples are 8 apples, so 4 times 2 thirds are 8 thirds. Namely,

$$4 \times \frac{2}{3} = \frac{4 \times 2}{3} = \frac{8}{3}.$$

The rule is that **multiplying a fraction by a whole number is done by multiplying its numerator by that number.** Another example:

$$6 \times \frac{3}{2} = \frac{6 \times 3}{2} = \frac{18}{2} = 9.$$

There is another way of viewing the same rule. We know that $\frac{2}{3} = 2 \div 3$. When the dividend in this expression, namely 2, is multiplied by 4, the first rule of change for division says that the quotient is multiplied by 4. Namely, $(4 \times 2) \div 3 = 4 \times (2 \div 3)$, which in the language of fractions reads: $\frac{4 \times 2}{3} = 4 \times \frac{2}{3}$. This is precisely the same rule as above.

Dividing a Fraction by a Whole Number

What is $\frac{2}{3} \div 4$? We know that $\frac{2}{3} = 2 \div 3$, and by the second rule of change for division, dividing the quotient $2 \div 3$ by 4 is tantamount to multiplying the divisor, 3, by 4. Namely, $(2 \div 3) \div 4 = 2 \div (3 \times 4) = 2 \div 12$. Returning to writing in fractions, this reads:

$$\frac{2}{3} \div 4 = \frac{2}{3 \times 4} = \frac{2}{12}.$$

The rule is: Dividing a fraction by a whole number is done by multiplying the denominator by the number. Another example:

$$\frac{5}{2} \div 6 = \frac{5}{2 \times 6} = \frac{5}{12}.$$

Here is a graphical illustration of the fact that $\frac{2}{3} \div 4 = \frac{2}{12}$.

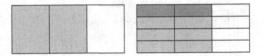

The grey area in the left drawing constitutes $\frac{2}{3}$ of the rectangle. In the drawing on the right, the horizontal lines cut the rectangle into 4 equal parts, hence the darkened area constitutes a quarter of these $\frac{2}{3}$. The two systems of lines, horizontal and vertical, partition the rectangle into 12 smaller equal rectangles, of which the 2 dark rectangles form $\frac{2}{12}$.

Equivalent Fractions

You better cut the pizza in four pieces because I'm not hungry enough to eat six.

Yogi Berra, Baseball Player

What happens if we multiply **both** the numerator and the denominator by 4? By the two rules we have just learned, the fraction first increases 4 times and then decreases 4 times! Like Alice in Wonderland after drinking 2 potions, it returns to its original size. When a number is multiplied by 4 and then divided by 4, it doesn't change, just as when 4 is added to a number and then subtracted from it, the number doesn't change.

$$10 \times 4 \div 4 = 10, \text{ just as } 10 + 4 - 4 = 10.$$

The conclusion is that replacing $\frac{2}{3}$ with $\frac{2 \times 4}{3 \times 4}$, that is, $\frac{8}{12}$, doesn't alter the number. The two fractions, $\frac{8}{12}$ and $\frac{2}{3}$, are equal. It is common terminology to call the two modes of writing "equivalent." We say that $\frac{8}{12}$ is the expansion of $\frac{2}{3}$ by 4.

Here is an example of the equality of $\frac{2 \times 4}{3 \times 4} = \frac{2}{3}$. In the right drawing, we divide each of the 2 colored thirds of the rectangle into 4 equal parts. The size of each small part is a quarter of a third: $\frac{1}{3 \times 4} = \frac{1}{12}$. Each third contains 4 such parts. Both thirds have 2×4, that is, 8 parts altogether. Therefore, two thirds are 8 parts of 12, as seen in the drawing:

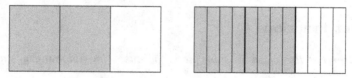

The same fraction can be written in many different ways. For example,

$$\frac{1}{2} = \frac{2}{4} = \frac{3}{6} = \frac{4}{8} = \cdots$$

(Here the fraction was expanded by 2, 3, 4, ...).

Summary

Expanding a fraction means multiplying its numerator and denominator by the same number. Expansion does not alter the value of the fraction.

Reduction

Question: *What is capitalism?*
Answer: *The exploitation of one man by another.*
Question: *And what is communism?*
Answer: *Vice versa.*

The reverse operation of expansion is called "reduction." It means dividing the numerator and denominator by the same number. Since expansion does not alter the fraction, the reverse operation, reduction, doesn't either. For example, in the fraction $\frac{8}{12}$, dividing the numerator and denominator by 4 results in $\frac{2}{3}$. Of course, we already know that $\frac{8}{12} = \frac{2}{3}$.

Summary

Reducing a fraction means dividing its numerator and denominator by the same number. Like expansion, reduction does not alter the value of the fraction.

Multiplying One Fraction by Another

Now we can go to the general case — multiplying two fractions. Take, for example, the product $\frac{4}{5} \times \frac{2}{3}$. Since $\frac{4}{5} = 4 \div 5$, multiplying by $\frac{4}{5}$ means multiplying by 4 and dividing by 5. By the rules we obtained for multiplying and dividing a fraction by a whole number, it follows that $\frac{4}{5} \times \frac{2}{3}$ is obtained from $\frac{2}{3}$ by first multiplying the numerator (namely, 2) by 4, and then multiplying the denominator (namely, 3) by 5. Thus:

$$\frac{4}{5} \times \frac{2}{3} = \frac{4 \times 2}{5 \times 3}.$$

The rule is: **When multiplying two fractions, the numerator is multiplied by the numerator and the denominator by the denominator.**
Another example:

$$\frac{3}{2} \times \frac{7}{6} = \frac{3 \times 7}{2 \times 6} = \frac{21}{12},$$

which, upon reduction by 3, is $\frac{7}{4}$.

Reduction Before Multiplication

In the last example we could have saved some effort by doing the reduction before the multiplication. Since we know that 3 is going to be a factor of the resulting numerator, and 6 a factor of the resulting denominator, we could reduce them in advance — dividing both by 3. Thus

$$\frac{3}{2} \times \frac{7}{6} = \frac{1}{2} \times \frac{7}{2} = \frac{1 \times 7}{2 \times 2} = \frac{7}{4},$$

which, if we wish, we can write as $1\frac{3}{4}$. This becomes significant when the numbers involved are large. For example, in the expression $\frac{27}{16} \times \frac{10}{9}$ we can (and should) reduce first by 9, by dividing both 27 and 9 by 9, and reduce by 2, by dividing both 16 and 10 by 2. This yields:

$$\frac{27}{16} \times \frac{10}{9} = \frac{3}{8} \times \frac{5}{1} = \frac{15}{8} = 1\frac{7}{8}.$$

Taking a Fraction of Something is Multiplying it by the Fraction

A fact about fractions which confuses children and grownups alike is that taking a fraction of something is multiplying it by the fraction. For example, by the definition of $\frac{2}{3}$, we know that $\frac{2}{3}$ of 60 apples is obtained by dividing the 60 apples into 3 equal parts, each containing 20 apples, and taking 2 of these parts, giving 40 apples. On the other hand,

$$\frac{2}{3} \times 60 = \frac{2 \times 60}{3} = \frac{120}{3} = 40$$

(in our case, 40 apples). But why is this so? The answer can be given on two levels, one technical and the other more profound. On the technical level, by the definition of "taking $\frac{2}{3}$" we know that it is done by dividing into 3 equal parts and taking 2 of them, which means dividing by 3 and multiplying by 2 (taking 2 times something is multiplying it by 2). On the other hand, since $\frac{2}{3} = 2 \div 3$, multiplying by $\frac{2}{3}$ is the same: It is multiplying by 2 and dividing by 3.

On a deeper level, we have seen that multiplication and counting are almost identical. Taking 5 objects is the same as multiplying the object by 5, both meaning the repetition of that object. Thus, for example, 5 tens is 5×10, both being the repetition of a 10 five times, and both yielding 50. This carries over to fractions. $\frac{2}{3}$ of a whole is the same as $\frac{2}{3}$ times that whole. Taking a fraction of it is the same as counting it, a fractional number of times. So, taking $\frac{2}{3}$ of a whole is the same as counting it $\frac{2}{3}$ times, which is the same as repeating it $\frac{2}{3}$ times, which is the same as multiplying it by $\frac{2}{3}$.

Example: In a class there are 32 students, $\frac{5}{8}$ of whom are girls. How many boys are there?

First solution: There are

$$\frac{5}{8} \times 32 = \frac{5}{1} \times 4$$

girls (we reduced by 4 before multiplication, which is really dividing 32 by 8), namely 20 girls. Therefore there are $32 - 20 = 12$ boys.

Second solution: If $\frac{5}{8}$ of the class are girls, the remaining, namely $\frac{3}{8}$, are boys. So, there are

$$\frac{3}{8} \times 32 = \frac{3}{1} \times 4 = 12 \text{ boys.}$$

Taking a Fraction of a Fraction

The fact that "taking a fraction of a whole" and "multiplying the whole by that fraction" are the same is true regardless of what constitutes the whole. In particular, it is also true when the whole is itself a fraction. For example, what is a $\frac{1}{3}$ of a $\frac{1}{2}$ of an apple? It is $\frac{1}{3} \times \frac{1}{2}$ of the apple, namely $\frac{1 \times 1}{2 \times 3} = \frac{1}{6}$ of the apple. This we could also do by remembering that a $\frac{1}{2}$ of the apple is obtained by dividing it into 2, and that a $\frac{1}{3}$ of it is obtained by dividing the half into 3, so altogether we divide by $2 \times 3 = 6$.

Example: What is $\frac{3}{4}$ of $\frac{5}{8}$? (Note that here $\frac{5}{8}$ is a "pure number," namely without denomination, meaning it could be $\frac{5}{8}$ of anything). Answer:

$$\frac{3 \times 5}{4 \times 8} = \frac{15}{32}.$$

Example: A girl tossed two coins 100 times. In $\frac{1}{2}$ of the tosses the first coin showed "heads," and in half of these tosses the second coin showed "heads." In how many tosses did both coins showed "heads"? Answer: in $\frac{1}{2} \times \frac{1}{2} = \frac{1}{4}$ of the tosses, namely in 25 tosses.

Problem: If the girl tossed three coins 1000 times, in half of which the first coin showed "heads," of which in half of the tosses the second coin showed "heads," of which in half of the cases the third coin showed "heads," in how many tosses all coins showed "heads"? Could you generalize?

Dividing by a Fraction

We now come to what is considered as the stumbling block of operations with fractions: dividing by a fraction. The rule everybody remembers is "invert and multiply," namely, dividing by $\frac{5}{8}$ is multiplying by $\frac{8}{5}$. But almost nobody understands why. Let us first state this rule more formally: **Dividing by a fraction is done by multiplying by its reciprocal.**

Here the "reciprocal" is obtained by exchanging numerator and denominator. In other words: **Dividing by a fraction is done by dividing by its numerator and multiplying by its denominator.**

Examples:

$$20 \div \frac{2}{3} = 20 \times \frac{3}{2} = 10 \times \frac{3}{1} = 30,$$

$$\frac{10}{7} \div \frac{5}{8} = \frac{10}{7} \times \frac{8}{5} = \frac{2}{7} \times \frac{8}{1} = \frac{16}{7}.$$

But why is this rule true? Let us exemplify it in the exercise $\frac{4}{5} \div \frac{2}{3}$. Our aim is to show that it equals $\frac{4 \times 3}{5 \times 2}$. Let us first calculate $\frac{4}{5} \div 2$. We know that $\frac{4}{5} = 4 \div 5$, and by the first rule of change for division, dividing $4 \div 5$ by 2 is the same as dividing 4 by 5×2. That is $(4 \div 5) \div 2 = 4 \div (5 \times 2)$, or in the language of fractions,

$$\frac{4}{5} \div 2 = \frac{4}{5 \times 2}.$$

Next, since $\frac{2}{3} = 2 \div 3$, by the second rule of change for division, dividing by $2 \div 3$ yields 3 times as much as the mere division by 2. Namely,

$$\frac{4}{5} \div (2 \div 3) = \left(\frac{4}{5} \div 2 \right) \times 3,$$

and since $\frac{4}{5} \div 2 = \frac{4}{5 \times 2}$, this is $\frac{4}{5 \times 2} \times 3$, which as we know is $\frac{4 \times 3}{5 \times 2}$, as claimed.

If this sounds too technical, it is best to keep in mind the following: A big numerator of the divisor makes it bigger, and a big denominator makes it smaller. Thus, the larger the numerator, the smaller is the result of the division; the larger the denominator, the larger the result. Hence it makes sense that we are multiplying by the denominator, and dividing by the numerator.

If this is still hard, I believe that the next chapter, which is an interlude about some teaching principles, will make it easier.

Summary of the Rules

1. Multiplying a fraction by 2 (for example) means multiplying its numerator by 2.
2. Dividing a fraction by 3 (for example) means multiplying its denominator by 3.
    (Both rules are derived from the rules of change in division, and from the fact that fractions are division).
3. Since $\frac{2}{3}$ are 2 divided by 3, multiplying by $\frac{2}{3}$ means multiplying by 2 and dividing by 3.
4. Division is the reverse operation of multiplication. Therefore, dividing by $\frac{2}{3}$ is the opposite of multiplying by it — it means multiplying by 3 and dividing by 2.

Dividing by a Fraction Smaller than 1 Increases the Number

Multiplying 9 by $\frac{2}{3}$ results in:

$$9 \times \frac{2}{3} = 9 \times 2 \div 3 = 18 \div 3 = 6.$$

The result is smaller than 9. This is no wonder — we multiplied by 2 and divided by 3, which is greater than 2. In contrast, dividing by $\frac{2}{3}$ means multiplying by 3 and dividing by 2. When a number is multiplied by a number greater than the one it is divided by, the number increases. For example, $6 \div \frac{2}{3} = 9$. (Can you relate this to the previous exercise?) In other words, the result is a number greater than 6, that is, 9.

There is, of course, a connection between the two. If multiplying by $\frac{2}{3}$ reduces, then the reverse operation, dividing by $\frac{2}{3}$, increases.

A Conversation on Division by Fractions

This chapter has two aims in mind: providing another opportunity for the teaching of division by fractions, and illustrating a few teaching principles. The explanation of the rule of division by a fraction is probably hard for children. Here is a way which is more suited to being taught in class. It is given in a form of a conversation with my daughter, Geffen. When she was in fourth grade, I took her for a walk. During that walk, which lasted less than an hour, I taught her how to divide a number by a fraction. This conversation is reproduced here quite accurately, and is used also as an opportunity for illustrating a few teaching principles. The principles involved in each stage appear parenthesized and italicized.

I: We have learned how to multiply by a fraction. Let us now see how to divide by a fraction. For example, how to calculate $10 \div \frac{2}{3}$. How do you think we should start?

G: We should ask a simpler question. (*Start from the simplest question possible. Also, share with the student the principles of sound thinking. The rule of starting from the simple should be explained to the students.*)

I: Right. Let us start then with dividing by a simpler fraction. What is the simplest fraction you can think of?

G: $\frac{1}{2}$.

I: Indeed, $\frac{1}{2}$ is a fraction we know and understand well. What is the simplest exercise you can think of in which we divide by $\frac{1}{2}$?

G: $1 \div \frac{1}{2}$. (*Let the student invent the problems.*)

I: Can you calculate that?

G: Yes, it is 2, since $1 \div 2 = \frac{1}{2}$. If $6 \div 2 = 3$ then $6 \div 3 = 2$, and likewise if $1 \div 2 = \frac{1}{2}$ then $1 \div \frac{1}{2} = 2$.

(*Well, this is smart. Not every child would see that. But what follows does not necessitate such insight.*)

I: Very nice! But here is another way, which I myself understand better. $6 \div 2 = 3$ because 2 goes into 6 three times. Do you remember what we call such division?

G: Yes, containment division. (*Use precise words, and distinguish fine points of meaning.*)

I: If $6 \div 2$ means how many times 2 goes into 6, what does $1 \div \frac{1}{2}$ mean?

G: How many times does $\frac{1}{2}$ go into 1?

I: And how many indeed?

G: $\frac{1}{2}$ goes into 1 two times. So $1 \div \frac{1}{2}$ is 2. (*Inadvertently, we followed here another teaching principle: Try to see the same thing from as many viewpoints as possible. We saw two ways of calculating $1 \div \frac{1}{2}$.*)

I: Could you tell me now what is $3 \div \frac{1}{2}$? (*Add one ingredient at a time.*)

G: Yes. $\frac{1}{2}$ goes two times into 1. Three is 3 ones, so $\frac{1}{2}$ goes $3 \times 2 =$ six times into 3. So $3 \div \frac{1}{2}$ is 6.

I: Right. And $4 \div \frac{1}{2}$?

G: 8.

I: And $5 \div \frac{1}{2}$? (*Stabilizing the knowledge, by exercising.*)

G: 10.

I: Could you tell me the rule?

G: Yes, dividing a number by $\frac{1}{2}$ multiplies it by 2.

Because each 1 in the number contains 2 halves. (*After experiencing a rule, formulate it in words.*)

Remark: In class, extensive exercising is needed at this point. How many halves of an apple are there in 5 apples? What is $5 \div \frac{1}{2}$? How many times does $\frac{1}{2}$ go into 10? What is $10 \div \frac{1}{2}$? 13 chocolate bars were divided between children, and each got half a bar. How many children were there? What is $13 \div \frac{1}{2}$?

I: Fine. Let us now divide by a third. What is $1 \div \frac{1}{3}$?

G: $\frac{1}{3}$ goes into 1 three times, so it is 3.

I: And what about $4 \div \frac{1}{3}$?

G: $\frac{1}{3}$ goes three times into 1, so it goes $4 \times 3 = 12$ times into 4.

I: What is the rule for dividing by $\frac{1}{3}$?

G: Dividing a number by $\frac{1}{3}$ is multiplying it by 3.

I: Great. So, we know that dividing by $\frac{1}{2}$ is multiplying by 2, and dividing by $\frac{1}{3}$ is done by multiplying by 3. How do you divide by $\frac{1}{4}$?

G: By multiplying by 4.

I: Yes, you've got the principle. Dividing by one over a number is multiplying by the number. (*This would be too hard for her to formulate.*) Let us now divide by $\frac{2}{3}$, which is what we started with. Let us first return for a moment to division by $\frac{1}{3}$. If in a party there were 10 cakes, and every child got a $\frac{1}{3}$ of a cake, how many kids were there?

G: $10 \div \frac{1}{3} = 30$; there were 30 kids.

I: And suppose now that each kid gets $\frac{2}{3}$ of a cake, instead of $\frac{1}{3}$, that is, two times more than before. For how many kids will the cake suffice?

G: Every kid now gets what 2 kids got before. So, there will be half of 30 kids, which is 15.

I: Right. What exercise did you do here?

G: $10 \div \frac{2}{3}$, because we found how many times $\frac{2}{3}$ goes into 10.

I: Right, and what operations did you do?

G: I multiplied the 10 by 3, and divided by 2.

I: So, what is the rule for dividing a number by $\frac{2}{3}$?

G: You take the number, multiply it by 3 and divide by 2.

I: And what would be the rule for dividing by $\frac{3}{4}$?

G: Multiply by 4, and divide by 3.

I: Very nice. And what is the rule for dividing by a general fraction?

G: You multiply by the denominator, and divide by the numerator.

Solved Exercises on Fractions

This book advocates systematic study, but it cannot itself be a substitute for a textbook. Hence, at least as far as exercises are concerned, it cannot supply a systematic array of problems which cover all topics. Still, since fractions are such a rich topic, a few more sample exercises are needed.

Problem (repeated taking away of fractions of a whole): The area of a park was 60 acres. $\frac{1}{6}$ of it was taken for the construction of a road. Then $\frac{1}{6}$ of the remaining area was taken for a zoo that was opened in it. How many acres remained in the park?

Solution: After $\frac{1}{6}$ was taken, $\frac{5}{6}$ of the park remained, and $\frac{5}{6}$ of 60 is $\frac{5}{6} \times 60$ acres. After another sixth of the area is taken, there remained $\frac{5}{6}$ of that, namely:

$$\frac{5}{6} \times \frac{5}{6} \times 60 = \frac{5 \times 5 \times 60}{6 \times 6} \text{acres.}$$

Reducing by 6, this is: $\frac{25 \times 10}{6}$ acres. Reducing again by 2 yields:

$$\frac{25 \times 5}{3} = \frac{125}{3} = 41\frac{2}{3} \text{acres.}$$

Problem (adding a fraction, and finding the original number from the result): Mary has a $\frac{1}{4}$ more marbles than Joe. How many marbles does Joe have if Mary has 45 marbles?

Solution: First, we have to understand what the problem says. The natural way to interpret it is that Mary has as many as Joe, plus a $\frac{1}{4}$ of what Joe has. This means that she has $1 + \frac{1}{4}$, namely $\frac{5}{4}$ the number of Joe's marbles. Thus, the question is: If $\frac{5}{4}$ of a number is 45, what is the number? We already learned how to do that: multiply 45 by $\frac{4}{5}$. Reducing by 5, we get:

$$\frac{4}{5} \times 45 = 4 \times 9 = 36.$$

(A quick reminder of one way to see why we should multiply by $\frac{4}{5}$. To get from Joe's number to Mary's we multiply by $\frac{5}{4}$, hence to go the other way around, from Mary's number to Joe's, we should do the opposite: divide by $\frac{5}{4}$, which is multiplying by $\frac{4}{5}$.)

Problem (*renaming fractions*): Fill in the missing numbers, denoted in each exercise by $\triangle$:

$$1. \ \frac{4}{5} = \frac{\triangle}{15}$$

$$2. \ \frac{2}{\triangle} = \frac{10}{60}$$

$$3. \ \frac{3}{\triangle} = \frac{4}{5}.$$

Solution 1: To pass from the left to the right, we multiplied the denominator by 3. (15, the denominator on the right, is 3 times 5, the denominator on the left.) To keep the fractions equal, we should also multiply the numerator by 3. Thus $\triangle = 4 \times 3 = 12$.

Solution 2: To pass from right to left, we divided the numerator by 5. Therefore, we should also divide the denominator by 5, that is: $\triangle = 60 \div 5 = 12$.

Solution 3: This exercise is different in that the denominator of the fraction on the left will come out not as an integer, but as a fraction itself. To go from the right to the left, we multiplied the numerator by $\frac{3}{4}$ (that is, $3 = \frac{3}{4} \times 4$). Hence, to obtain $\triangle$, we should multiply 5, the denominator on the right, by $\frac{3}{4}$. This yields:

$$\triangle = 5 \times \frac{3}{4} = \frac{15}{4} = 3\frac{3}{4}.$$

Equation 3 then looks a bit strange:

$$\frac{3}{3\frac{3}{4}} = \frac{4}{5}.$$

The Common Denominator

Child: *Father, can you help me find the common denominator?*
Father: *Hasn't it been found yet? They've been looking for it since I was a child!*

A Common Language

Gerald Ford, a United States president, was prone to stumbling: he fell on more than one occasion in full view of the TV cameras. One of the jokes often told about him was that he wasn't able to chew gum and walk at the same time.

It is common to hear, concerning someone considered clumsy or stupid, that he *can't walk and chew gum at the same time*. In fact, it is quite difficult to perform two different types of operations simultaneously. This is exactly what happens when trying to add fractions. A fraction is actually division. Therefore, adding fractions involves two operations: division and addition. A mixture of sorts. No wonder it's complicated. It requires what has become one of the symbols of elementary school mathematics: a common denominator.

In the chapter *Conveying Abstractions*, we likened the common denominator to a common language. When the fraction $\frac{1}{5}$ meets the fraction $\frac{2}{5}$ they can talk. They live in the same world, the world of fifths, and therefore they have no problem connecting. Just as 1 apple plus 2 apples are 3 apples, so one fifth plus two fifths are three fifths: $\frac{1}{5} + \frac{2}{5} = \frac{3}{5}$.

As in real life, the trouble begins when the two fractions don't speak the same language, that is, when they don't have the same denominator. In this case there is no other choice but to find for them a common language, which means giving them the same denominator, that is, a common denominator. In life this isn't always possible, but in arithmetic it is.

The Pizzeria Owner and the Indecisive Customer

I like to demonstrate the principle of the common denominator using a story about a pizzeria owner and his indecisive customer.

A customer ordered pizza for his 2 children, and asked that it be divided into 2 equal parts. He suddenly remembered that a friend might join them.

Divide the pizza, he asked, into equal parts, so that I can divide it between 2 children or 3 children. Into how many parts will the baker divide the pizza?

The answer is, of course, 6 parts. If the pizza is to be shared by 2 children, each child can receive 3 parts (one can say that $\frac{1}{2}$ is $\frac{3}{6}$), while if a third child arrives to share it, each child would get 2 parts ($\frac{1}{3} = \frac{2}{6}$).

By the way, dividing into 2 and 3 is not only the simplest example, it is also the most familiar, since pizzas are indeed often divided into 6 parts.

On the second day, the customer again ordered pizza and said, "Today 3 children will probably be sharing the pizza. But another child might join them and then there would be 4." Into how many parts should the pizza be divided? 12, of course. If there are 3 children, each will receive 4 slices, and if there are 4 children, each will receive 3 slices.

This story should be repeated again and again, for an entire lesson or even longer. With 2 children and 5, with 2 and 4 (which is a different case, since 4 is divisible by 2), with 4 and 6.

Common Multiples

Why is it that when a pizza is divided into 6 equal parts, both 2 children and 3 children can share it equally? The answer is simple: because 6 is divisible both by 2 and by 3.

Another way to say the same is that 6 is a **common multiple of 2 and 3**. Being divisible by a number means being a multiple of it: 6 is a multiple of 2, since it is 3 times 2, and also a multiple of 3, since it is 2 times 3.

However, if 6 is divisible both by 2 and by 3, then we have found the required common language for $\frac{1}{2}$ (from dividing by 2) and $\frac{1}{3}$ (from dividing by 3): the language of sixths. We've discovered that both can be expressed as sixths: $\frac{1}{2} = \frac{3}{6}$ and $\frac{1}{3} = \frac{2}{6}$.

The term used for this is "common denominator" — 6 is a common denominator of $\frac{1}{2}$ and $\frac{1}{3}$. This is the third way of saying what we have already expressed in two different ways: 6 is divisible by 2 and by 3, and 6 is a common multiple of 2 and 3.

Now we can finally perform the addition $\frac{1}{2} + \frac{1}{3}$. Both $\frac{1}{2}$ and $\frac{1}{3}$ can be expressed as sixths: $\frac{1}{2} = \frac{3}{6}$ and $\frac{1}{3} = \frac{2}{6}$, and therefore, they can be added:

$$\frac{1}{2} + \frac{1}{3} = \frac{3}{6} + \frac{2}{6} = \frac{5}{6}.$$

The Trick: Expansion

Writing $\frac{1}{2} = \frac{3}{6}$ means expanding the fraction $\frac{1}{2}$ by multiplying both the numerator and the denominator by 3. $\frac{1}{3} = \frac{2}{6}$ is also a result of an expansion: multiplying the numerator and denominator of $\frac{1}{3}$ by 2.

Finding a common denominator, therefore, requires expansion of both added fractions, so that they have the same denominator.

Here is another example, this time without the aid of the pizzeria owner: calculating $\frac{2}{3} + \frac{1}{4}$.

First a common multiple of both denominators, that is, of 3 and 4, must be found. At this stage, we will guess: 12 is divisible both by 3 and by 4. (Can you discover, based on this example and maybe on the previous one, a method for selecting a common multiple?)

Our aim is now to expand the fraction $\frac{2}{3}$ so that its denominator will be 12. The denominator, 3, must be multiplied by 4 to yield 12, and therefore, the numerator must also be multiplied by 4, namely $\frac{2}{3} = \frac{8}{12}$.

The $\frac{1}{4}$ also needs to be expanded to a denominator of 12. This requires multiplying the denominator by 3. Therefore, the numerator must also be multiplied by 3, namely $\frac{1}{4} = \frac{3}{12}$. The result is thus:

$$\frac{2}{3} + \frac{1}{4} = \frac{8}{12} + \frac{3}{12} = \frac{11}{12}.$$

The Simplest Common Multiple

The numbers 3 and 4 have many common multiples. For example, 120 is one of their common multiples. However, among all common multiples there is one that is simplest, or most natural: *the product of the two numbers*, that is, 3×4. For 3×4 is obviously a multiple both of 3 and of 4! In fact, this is the first common multiple we guessed, 12!

The rule is: The product of the denominators is always their common multiple (in other words, common denominator).

Summary

The product of the denominators is a common denominator.

Summarizing Formula: Cross-Multiplying and Adding

The addition of fractions has a simple summarizing formula. In order to study it, we will calculate another example: $\frac{2}{5} + \frac{1}{3}$. The common denominator is the product of the denominators: $5 \times 3 = 15$. To bring $\frac{2}{5}$ to this denominator, $\frac{2}{5}$ must be expanded by the denominator of the other fraction, that is, by 3. The numerator will then be 2×3. To bring $\frac{1}{3}$ to a denominator of 15, $\frac{1}{3}$ must be expanded by the denominator of the other fraction, that is, by 5, and the numerator will then be 1×5. When summing the fractions, the numerator is $2 \times 3 + 1 \times 5$. This is the sum of products of crossed numbers: the numerator of one fraction times the denominator of the other, plus the numerator of the second fraction times the denominator of the first. The result is:

$$\frac{2}{5} + \frac{1}{3} = \frac{2 \times 3 + 1 \times 5}{5 \times 3}.$$

This rule can be phrased generally: The sum of two fractions has, as a denominator the product of the fractions' denominators, and as a numerator the sum of two products of the numerators cross-multiplied by the denominators, namely each numerator is multiplied by the denominator of the other fraction.

Subtracting Fractions

We now know how to add fractions. What about subtraction? It is exactly the same. For example:

$$\frac{2}{5} - \frac{1}{3} = \frac{2 \times 3}{5 \times 3} - \frac{1 \times 5}{3 \times 5} = \frac{2 \times 3 - 1 \times 5}{15} = \frac{6 - 5}{15} = \frac{1}{15}.$$

An Additional Advantage of the Common Denominator:

Comparison of Fractions

Which fraction is bigger, $\frac{2}{3}$ or $\frac{4}{5}$? This question has an easy answer: $\frac{2}{3}$ is 1 minus $\frac{1}{3}$, while $\frac{4}{5}$ is 1 minus $\frac{1}{5}$. $\frac{1}{5}$ is smaller than $\frac{1}{3}$, and therefore, 1 minus $\frac{1}{5}$ is bigger than 1 minus $\frac{1}{3}$. In other words, $\frac{2}{3}$ is smaller than $\frac{4}{5}$.

   However, comparison questions can be more complicated: Which is bigger, $\frac{2}{7}$ or $\frac{5}{17}$?

One option is to find a common denominator for the two fractions.
$$\frac{2}{7} = \frac{2 \times 17}{7 \times 17} = \frac{34}{119}, \text{ while } \frac{5}{17} = \frac{5 \times 7}{17 \times 7} = \frac{35}{119}.$$

Therefore, $\frac{5}{17}$ is bigger, just as 35 apples are more than 34 apples. When two fractions have a common denominator, the fraction with the bigger numerator is the larger of the two.

The Least Common Denominator

Is it Really That Bad?

In the previous chapter we learned that when adding and subtracting fractions, the product of their denominators is a common denominator. Those who remember the common denominator from their school days are probably surprised. Is this it? Just multiplying the denominators? But the common denominator is infamous! Generations of students shudder to hear its name!

The reason is that there is another little twist. Sometimes it is the lowest common denominator that is expected, that is, the smallest common multiple of the denominators. While there are good reasons to ask for a low common denominator, it complicates the calculations a bit.

For example, in the exercise $\frac{1}{4} + \frac{1}{6}$, a common multiple of 4 and 6 is required. In the previous chapter we learned how to find one common multiple by multiplying the denominators by each other. In this case, the common multiple is 4×6, which is 24. However, there is a smaller common multiple: 12. 12 is also divisible both by 4 and by 6, and so can also serve as a common denominator.

There are a few reasons for searching for the lowest common denominator, and for teaching it. One reason is economy in calculation. To calculate the exercise the fractions must be expanded, and the lower the common denominator, the smaller the factors of expansion. The other reason is educational: it demands from the student a better understanding of fractions. A student who, in the exercise $\frac{1}{100} + \frac{1}{200}$, takes as a common denominator the product of 100 and 200, namely 20,000, doesn't really understand the spirit of what he is doing, and performs the calculation mechanically. Still another reason is that finding the lowest common denominator teaches certain principles of thought, in particular, principles associated with breaking numbers into factors, and, as we will see, classification. But my belief is that the topic of the smallest common denominator should be taught separately from the topic of common denominator *per se*, and that the distinction should be declared to the student. Divide and conquer!

Three Types of Fraction Pairs

As mentioned, one of the principles the least common denominator teaches is *classification*. Fraction pairs can conveniently be categorized into three types:

Type A: The denominators are "relatively prime," which means that there is no number, aside from 1, which divides them both. For example: $\frac{1}{3} + \frac{1}{4}$.

In this case the least common denominator is the product 3×4, namely, 12. There is no room for economy.

Type B: One of the denominators is divisible by the other. For example: $\frac{1}{4} + \frac{5}{12}$. 12 is divisible by 4. This case is also simple: The least common denominator is the larger of the two denominators. In our example, it is 12, for it is divisible both by itself and by 4. The fraction $\frac{1}{4}$ must be expanded by 3; the result is $\frac{3}{12}$. The sum is, therefore, $\frac{3}{12} + \frac{5}{12}$, which is $\frac{8}{12}$.

Type C: The two denominators have a common divisor, that is, a number greater than 1 that divides them both, but they are not divisible by each other. For example, in $\frac{1}{4} + \frac{1}{6}$, both denominators are divisible by 2, but 6 is not divisible by 4. This is the most interesting case, and the only one that poses a difficulty.

To deal with this case, we need to learn a new term: the *prime factors* of a number. These are the prime numbers whose product is the given number. (Remember that a number is prime if it has no divisor apart from itself and 1. For example, 7 is prime but 6 is not since it is divisible by 2 and 3.) A factor may appear more than once, that is, repeat itself. For example, $24 = 2 \times 2 \times 2 \times 3$, and so the factors of 24 are 2, 2, 2 and 3.

Let's return to the example $\frac{1}{4} + \frac{1}{6}$. We are searching for a number that is divisible both by 6 and by 4. Let's give it a try. Begin with one of the numbers, 6. Obviously , it fulfills one of the two conditions: It is divisible by 6, that is, by itself. What is missing is the second condition, being divisible by 4. The reason why 6 is not divisible by 4 is that it doesn't contain all of the factors contained in 4. 4 includes the factor 2 twice, since $4 = 2 \times 2$. 6, on the other hand, contains the factor 2 only once, since $6 = 2 \times 3$. In order to be divisible by 4, 6 is missing one copy of the factor 2. To fix the problem, 6 should be multiplied by 2. The result, 12, contains the factor 2 twice, for $12 = 2 \times 2 \times 3$. This means that 12 is divisible by 4, and it is indeed the least common multiple — the required common denominator.

The rule is:

To find the least common multiple of two numbers, A and B, multiply A by those prime factors of B missing in A.

Obviously, either one of the two numbers can be selected as a starting point. Here's another example: Find the least common multiple of 10 and 15. Begin with one of the numbers, for instance, 10. Try to multiply it so that the resulting number will also be divisible by 15. The factors of 15 are 3 and 5, and the common multiple must contain them. However, the factor 5 is already contained in 10 (10 is divisible by 5), and therefore, there is no need to multiply by it. The second factor, 3, on the other hand, is not contained in 10. Therefore, 10 should be multiplied by it. The result, 30, is the least common multiple.

As a last example, let's calculate $\frac{3}{10} + \frac{1}{8}$. Select one of the denominators, say 10, and multiply it by the factors contained in 8 but not in 10. The factorization of 8 is: $8 = 2 \times 2 \times 2$, so 8 contains 3 factors of 2. In 10, which equals 2×5, there is only one factor 2. Therefore, 8 contains two more factors of 2. Hence, to become divisible by 8, 10 must be multiplied by 2 twice, that is, by 4. The resulting number, 40, is the least common denominator.

To perform the addition, expand both fractions in the sum so that their denominator is equal to the common denominator. In this case: $\frac{3}{10} = \frac{12}{40}$, $\frac{1}{8} = \frac{5}{40}$. Now, addition is possible: $\frac{12}{40} + \frac{5}{40} = \frac{17}{40}$.

Mixed Numbers

Thou shall not plow with an ox and an ass together.

Deuteronomy 22:10

What is a Mixed Number?

A mixed number is the sum of a natural number and a fraction. For example, $7 + \frac{4}{9}$, or as conventionally written: $7\frac{4}{9}$.

Mixed numbers arise in various contexts. For instance, when adding fractions, the sum $\frac{3}{4} + \frac{3}{4}$ equals $1\frac{1}{2}$. Or: $17 \div 7$ is $2\frac{3}{7}$ (7 is contained 2 times in 17, leaving a remainder of 3, which needs to be divided by 7).

Improper Fractions

"I have just been thinking and I have come to a very important decision," said Pooh. "These are the wrong sort of bees." "Are they?" said Christopher Robin. "Quite the wrong sort. So I should think they would make the wrong sort of honey, shouldn't they?"

A.A. Milne, **Winnie the Pooh**

The expression $\frac{17}{7}$ is called an "improper fraction." Why? Because it is only posing as a fraction. In fractions, as defined in elementary school, the numerator must be smaller than the denominator (professional mathematicians are the ones who don't make this distinction, and call $\frac{17}{7}$ a fraction, too.)

As a matter of fact, there is nothing improper about an improper fraction. It adds and subtracts as a fraction, and multiplies and divides as a fraction.

How Does an Improper Fraction Become a Mixed Number?

The first answer should be: there is no need! Let it be! An improper fraction can remain improper.

The only disadvantage of improper fractions is that it is difficult to estimate their size. Estimating $\frac{17}{7}$ requires some thought. When it is written as $2\frac{3}{7}$ it is easy to place between 2 and 3.

How is $\frac{17}{7}$ turned into a mixed number? The fraction line represents division. If so, divide! The result of the division $17 \div 7$ is $2\frac{3}{7}$.

Summary

To turn an improper fraction into a mixed number, divide the numerator by the denominator.

How are Mixed Numbers Added?

The principle is very similar to that of adding decimal numbers. The whole numbers and the fractions are added separately, and if a whole number can be grouped from the sum of the fractions, then the result is transferred to the wholes. Here's an example: $1\frac{3}{4} + 3\frac{1}{2}$.

$$1\frac{3}{4} + 3\frac{1}{2} = 1 + 3 + \frac{3}{4} + \frac{1}{2} = 4 + \frac{3}{4} + \frac{1}{2} = 4 + \frac{5}{4} = 4 + 1\frac{1}{4} = 5\frac{1}{4}.$$

How are Mixed Numbers Subtracted?

Here, too, the principle is similar to that of subtracting decimal numbers.

Let's begin with a simple example: $3\frac{5}{7} - 1\frac{2}{7}$. In decimal numbers, the tens are subtracted from the tens and the ones from the ones. Here, the wholes are subtracted from the wholes and the fractions from the fractions. The result is 3 wholes minus 1 whole, which are 2 wholes; and $\frac{5}{7} - \frac{2}{7}$, which equals $\frac{3}{7}$. Altogether: $2\frac{3}{7}$.

As in decimal numbers, matters get complicated when a number needs to be exchanged for smaller units — in this case, fractions. Take for instance, $3 - \frac{2}{7}$. The 3 doesn't have enough "fraction parts" from which to subtract $\frac{2}{7}$. What should we do? Get change! Take 1 from the 3 and exchange it for sevenths: $1 = \frac{7}{7}$. From the 7 sevenths, 2 sevenths can be subtracted:

$$3 - \frac{2}{7} = 2 + 1 - \frac{2}{7} = 2 + \frac{5}{7} = 2\frac{5}{7}.$$

Here is a more complex example: $5\frac{1}{4} - 3\frac{5}{6}$.

The problem here is that $\frac{1}{4}$ is smaller than $\frac{5}{6}$. If we try to calculate $\frac{1}{4} - \frac{5}{6}$, the result will be a negative number. It won't be the end of the world, but we haven't learned about negative numbers yet.

Therefore, we change again. Of the 5 wholes in the minuend, one is changed to $\frac{4}{4}$. Of the 5, only 4 wholes remain and the $\frac{1}{4}$, with the addition of $\frac{4}{4}$, becomes $\frac{5}{4}$.

We now have:

$$4 + \frac{5}{4} - 3\frac{5}{6} = 1 + \frac{5}{4} - \frac{5}{6}.$$

All that is left to calculate is $\frac{5}{4} - \frac{5}{6}$. The least common denominator is 12, and therefore the first fraction must be expanded by 3 and the second by 2, resulting in:

$$\frac{5 \times 3}{4 \times 3} - \frac{5 \times 2}{6 \times 2} = \frac{15}{12} - \frac{10}{12} = \frac{5}{12}.$$

Remember that we need to calculate $1 + \frac{5}{4} - \frac{5}{6}$. Since $\frac{5}{4} - \frac{5}{6} = \frac{5}{12}$, the result is $1\frac{5}{12}$.

Transforming Mixed Numbers into Improper Fractions

As mentioned, there is no special reason to transform improper fractions into mixed numbers. But the other way around, transforming mixed numbers into improper fractions, is often necessary. It is required for multiplication and division. Let's use an example to see how it's done. How is a number such as $3\frac{2}{7}$ turned into an improper fraction? The 3 can be expressed as a fraction with a denominator of 7. The number 1 is 7 sevenths. Thus 3, which is 3 ones, is 3 times 7 sevenths, namely $\frac{21}{7}$. Therefore,

$$3\frac{2}{7} = \frac{21}{7} + \frac{2}{7} = \frac{23}{7},$$

which is the required translation of the number into an improper fraction.

Examining what we just did, we discover a simple rule: The numerator, 23, is obtained by multiplying the integer part, 3, of $3\frac{2}{7}$, by the denominator 7, and then adding 2. Using this rule to transform $5\frac{1}{3}$ we see:

$$5\frac{1}{3} = \frac{5 \times 3 + 1}{3} = \frac{16}{3}.$$

How are Mixed Numbers Multiplied?

Now we can multiply and divide mixed numbers. Take, for example, an exercise such as $1\frac{1}{2} \times 2\frac{1}{3}$. We've already mentioned that fractions like multiplication and division, and multiplication and division like them.

Therefore, mixed numbers should be turned into fractions, even if it means improper fractions!

$1\frac{1}{2} = \frac{3}{2}$ and $2\frac{1}{3} = \frac{7}{3}$. Therefore, $1\frac{1}{2} \times 2\frac{1}{3}$ is the same as

$$\frac{3}{2} \times \frac{7}{3} = \frac{3 \times 7}{2 \times 3} = \frac{7}{2} = 3\frac{1}{2}.$$

How are Mixed Numbers Divided?

Danny and Joe go to a restaurant. Danny orders coffee with sugar and milk. Joe asks for the same, only with tea instead of coffee, sweet 'n' lo instead of sugar, and lemon instead of milk.

How are mixed numbers divided? Well, just as they are multiplied. But, unlike the joke, it is really the same.

For example, to perform the division exercise $1\frac{1}{2} \div 2\frac{1}{3}$, the mixed numbers are turned into improper fractions. We performed this transition in the previous exercise, so we can write it down straight away:

$$\frac{3}{2} \div \frac{7}{3} = \frac{3 \times 3}{2 \times 7} = \frac{9}{14}.$$

D. Decimals

Decimal fractions are the result of an encounter between fractions and the decimal system. They are based on the fact that the place value system can be continued on into fractions. On the left of the ones digit are the tens, hundreds, thousands, etc. On the right are the fractions: tenths, hundredths, thousandths, and so forth.

In spirit, in their calculation methods and in the ideas required to understand them, decimal fractions belong to the realm of the decimal representation of numbers more than to that of fractions.

Decimal fractions are usually taught in the fifth or sixth grade, but they can be initially introduced much earlier, through money. For example, writing "3 dollars and 27 cents" as "3.27 dollars."

Addition and subtraction in decimal fractions are as simple as addition and subtraction in natural numbers. Multiplication and division are performed by ignoring (just for a moment!) the decimal point, performing the calculation as if they were ordinary numbers, and then returning the decimal point to its proper place.

The decimal method of writing fractions has one disadvantage: Not all fractions can be written as decimal fractions. Only fractions that include 2 and 5 alone as factors of their denominators can be written as decimal fractions. But every fraction can be approximated as closely as one wishes by a decimal fraction.

Decimal Fractions

Professor Elemeno Invents Decimal Fractions

As a sequel to the story about the invention of the decimal system by King Krishna, here is another fictional story about the invention of decimal fractions. Or to be precise, their rediscovery. This is a story for grownups as it's a little too complex for children.

* * *

Professor Elemeno arrived one day at his house on Exwhyzee St., with sheer excitement. "Anna-Lisa," he said, his face blazing. "I invented a wonderful thing today. Something no one has thought of before."

Anna-Lisa, accustomed to her husband's inventions, wasn't particularly moved. "Yes," she said, whipping up 3 eggs for a cake. "What is it this time?"

"Look," said Elemeno. "In the number 354, the leftmost digit, 3, tells us how many hundreds there are, the second from the left, 5, how many tens, and the third, 4, how many ones. 100, 10, 1 — ten times less each time! Why not continue the same way? Ten times less than 1 is a tenth. If I put a digit on the right, it will represent tenths! For example, when I write 3547, I mean 3 hundreds, 5 tens, 4 ones and 7 tenths! Isn't it ingenious?"

"Maybe," said Anna-Lisa, pouring 2.3 cups of flour into the mixture, as instructed by the recipe. "But you are the only one who will know what you mean. When I read the number 3547, I see three thousand five hundred forty seven. How do you distinguish between the two numbers?"

Professor Elemeno thought long and hard. What is wrong here? In 354 the 4 is 4 ones, so why doesn't the 7 he adds on the right represent $\frac{7}{10}$?

"It's because the entire number moves to the left," said Anna-Lisa, while turning on the mixer so that Elemeno could hardly hear her. "Now the 7 represents ones. The 4 moved to the left and represents tens," she explained. "You need something that will prevent the 4 from being pushed to the left. For instance, a barrier sign |. Write 354|7 and everyone will know that on the left of the barrier are the ones and on the right are the tenths."

"Brilliant," said Elemeno. "Why didn't I think of that myself?"

Anna-Lisa lapsed into thought. She took a piece of paper and wrote: 354|7. "Good, good. Not for nothing did I marry a mathematician," she said to herself. "He does have interesting ideas now and then. But the barrier, |, is too much like the number 1. It may be confusing. Maybe we should put an interesting drawing instead? For instance, ♡? We'll write: 354♡7."

"It's too difficult to draw," said Elemeno. "Why not just put a dot there instead? Then it will look like this: 354.7. It's pretty clear, and easy to draw. And now if I write 354.78, everyone will know that the 7 means tenths and the 8 means hundredths."

And so the Elemeno family invented decimal fractions. It is true that these fractions have been known in Europe for over 500 years. But let's not disillusion Elemeno and Anna-Lisa.

Summary

Decimal fractions are a continuance into fractions of the place value system. To the left of the decimal point the digits represent ones, tens, hundreds, and so forth. To the right of it, they represent tenths, hundredths, thousandths, and so forth.

The story about Professor Elemeno may be fictional, but it is mathematically correct. Decimal fractions are a natural continuance of the decimal system. When moving to the right, the digits go from representing thousands to representing hundreds, then tens and ones, each time 10 times less. Beyond the ones, there are the tenths, and then the hundredths. But, as Anna-Lisa pointed out, the place of switching from ones to tenths must be marked. Otherwise, the entire number will be "pushed" to the left. This separation is achieved by the decimal point.

Historical Note

Who Really Invented the Decimal Fractions?

As with the representation of numbers, the idea behind decimal fractions, continuing the number beyond the decimal point, was invented by the Babylonians. As mentioned earlier, they collected groups of 60 items, and therefore the value of the digit following the "point" was $\frac{1}{60}$ and the value of the next digit was $\frac{1}{3600}$. However, they didn't mark the point. They understood its location from the context.

Surprisingly, even after they started using the decimal system to represent ordinary numbers, the Europeans continued to use the "60" method in fractions up until the fifteenth century! The decimal fractions system as it is used nowadays is only 500 years old.

A relic of the Babylonian system is left in the breaking up of the hour into 60 minutes, and of the minute into 60 seconds. A similar system is preserved in the measuring of angles, where the units are also called "minutes" and "seconds."

The Advantage of Decimal Fractions

The major advantage of decimal fractions is that they are convenient for calculations. They bring with them the ready-made mechanism of the decimal system, with its familiar algorithms of calculation. This means that they also include a ready-made common denominator. Want to add 2.3 and 4.5? No problem: The 2 and the 4 both represent ones, which means they have a common denomination. The 3 and the 5 represent tenths, so they also have a common denomination.

It is also easy to estimate the size of a decimal fraction, and compare the sizes of fractions.

Infinite Decimal Fractions

Nothing is perfect. Decimal fractions must have some flaw, and indeed they do. Not every fraction can be written as a decimal fraction, at least not in a finite form. Some fractions — in a certain sense, most of them — can be represented decimally only as infinite decimal fractions.

The most familiar of all may be the representation of the fraction $\frac{1}{3}$. In decimal writing it is $0.333\ldots$, with the three dots meaning that the series of 3s goes on indefinitely.

The meaning of the equality $0.333\ldots = \frac{1}{3}$ is that the numbers, $0.3, 0.33, 0.333, \ldots$, draw nearer and nearer to $\frac{1}{3}$. The accurate terminology is that their distance from $\frac{1}{3}$ tends to 0.

An even simpler example is the infinite decimal fraction $0.999\ldots$. It is actually equal to 1. Why? Because the distance from 0.9 to 1 is $\frac{1}{10}$. The distance from 0.99 to 1 is $\frac{1}{100}$. The distance from 0.999 to 1 is $\frac{1}{1000}$. The distance to 1 tends to 0, meaning that 1 is the limit. It turns out that a number can have two different decimal representations: $1 = 0.999\ldots$. A little confusing, but that's the way it goes.

Why is it Permissible to Adjoin a 0 to the Right of a Decimal Fraction?

Adjoining a 0 to the right of an ordinary decimal number alters it. For example, adjoining a 0 to the right of the number 58 turns it into 580. In decimal fractions, in contrast, adjoining a 0 does not alter the number. For example, 5.8 = 5.80.

Why is this so? Adjoining a 0 to an ordinary number pushes the entire number to the left. The digit 8 in the number 58, for example, which previously represented ones, represents the tens of the number 580 after the addition of a 0.

In decimal fractions, the decimal point prevents the shift. It defines the role of each digit, and therefore, there is no danger in adjoining a 0. For instance, the 8 in 5.80 represents tenths, just as in 5.8.

12.0 = 12

Take a look at the number 12.0. To the left of the 0 are ones, just as in 12. Therefore, the two numbers are equal. Putting a decimal point at the end of a number doesn't alter it.

Is it permissible to just place a decimal point, without a 0, and write 12.? Of course. There is no need to place the 0 to the right of the decimal point, just as there is no need to write 12.10 instead of 12.1, even though it is true that there are 0 hundredths. If nothing is written, it is obvious there is a 0. The problem is that the point can be confused with a period at the end of a sentence. Therefore, it is customary to include the 0, and write 12.0.

For the same reason, it is also permissible to write ".34" instead of "0.34"; the decimal point clarifies that there are 0 ones, and so it is not necessary to write the 0.

The Power Operation

Decimal fractions can first be introduced through money: a second-grader can understand the meaning of "3.25 dollars," which he will read as "3 dollars and twenty five cents." A third-grader can understand "13.25 dollars." Formal teaching, with the name "decimal fraction" and the meaning of the decimal point, is included in the fourth, fifth or sixth grade. In these grades, as one of the first steps to understanding decimal fractions, the comprehension of the decimal system can be enhanced by learning the power operation.

In the higher grades, the decimal fractions can be linked to the *power operation*. The digits of a decimal number represent 1s, then 10s, 100s, and so forth. Ones, tens, hundreds, thousands ... What are all these? Each number in the series 1, 10, 100, 1000 is a result of the multiplication of its predecessor by 10. Such an operation, repeated multiplication, is called "power." In this case, powers of 10. For example, 1000 is 10 to the power of 3, since it is a product in which 10 appears 3 times: $1000 = 10 \times 10 \times 10$. It is denoted as $1000 = 10^3$.

One hundred is 10 to the power of 2, or 10^2, since it is 10×10, a product where 10 appears twice. Ten is 10^1, since it is a product where 10 appears once.

How do we know what power of 10 lies before us? Simply count the zeros. In 1000, which is 10 to the power of 3, there are 3 zeros.

How are Powers Multiplied?

What is $10^2 \times 10^4$?

10^2 means 10×10, and 10^4 means $10 \times 10 \times 10 \times 10$. Their product is therefore $(10 \times 10) \times (10 \times 10 \times 10 \times 10)$, in which 10 appears $2 + 4$ times, that is, 6 times. Conclusion: $10^2 \times 10^4 = 10^6$, that is, 10 to the power of the sum of the two powers.

Summary

A power represents repeated multiplication. For example, 3 to the power of 4 is $3 \times 3 \times 3 \times 3$, a product which includes 3 four times. It is marked as 3^4. The name "power" originates from the fact that this is indeed a powerful operation: Its result can be large, even when the numbers involved are small!

What is 10^0?

We have seen that $10^3 = 1000$, $10^2 = 100$ and $10^1 = 10$. From left to right, each number is a tenth of its predecessor. Therefore, if we continue this series to the right, the next term will be a tenth of 10, which is 1. Thus, $10^0 = 1$.

This makes sense. We said that the power is determined by how many zeros there are to the right of the digit 1, so 10 to the power zero is 1 with no zeros to its right. Yet another way to see it is this: What is 10 to the power of 2? It is the number 1 multiplied by 10 twice. That is, $10^2 = 1 \times 10 \times 10$. What is 10 to the power of 0? It is the number 1 multiplied by 10 zero times. Do not confuse this with multiplying by 0. There is simply no multiplication, nothing is done. It is left as is, namely as 1.

Negative Powers

Let's write the powers of 10: $10^0 = 1$, $10^1 = 10$, $10^2 = 100$, $10^3 = 1000, \ldots$ Now, let's read the series of exponents (powers) from right to left and continue beyond the 0, like this: $3, 2, 1, 0, -1, -2, -3, \ldots$ If so, the next powers are negative powers:

$$10^3 = 1000, \ 10^2 = 100, \ 10^1 = 10, \ 10^0 = 1,$$
$$10^{-1} = \frac{1}{10}, \ 10^{-2} = \frac{1}{100}, \ 10^{-3} = \frac{1}{1000}, \ \ldots$$

In decimal fractions, the first digit on the right of the decimal point represents, as recalled, tenths, that is, 10^{-1}. The second digit represents hundredths, that is, 10^{-2}, and so forth. The digits on the right of the decimal point represent *negative* powers of 10, while the digits to the left of the decimal point represent *non-negative* powers of 10.

Decimal fractions are, therefore, an expansion of the decimal system to negative powers.

Calculation in Decimal Fractions

What Happens to the Decimal Fraction When the Decimal Point is Shifted to Either Side?

In this chapter we will study two subjects: First, how to express a decimal fraction as an ordinary fraction and vice versa; second, how to calculate arithmetical operations with decimal fractions. Both require knowledge of a simple principle: What happens to the decimal fraction when its decimal point is shifted to either side?

Let's take a look at an example: the number 57.34. What will happen if we shift the decimal point one place to the right? The result will be 573.4. The 5, which previously enumerated tens, now enumerates hundreds (5 hundreds). In other words, it represents a number 10 times greater than before. The same happened to the 7 — it previously enumerated ones and now tens, also 10 times greater. The 3, which previously enumerated tenths, now enumerates ones — also ten times greater.

Each digit now represents 10 times the number it represented before. In other words, the entire number increased 10 times.

Conversely, shifting the decimal point to the left reduces the number 10 times. For instance, 0.75 is a tenth of 7.5.

And what happens if the point is shifted 2 places to the right? Shifting it one place increases the number 10 times; one more place also increases it 10 times. Altogether, the number is multiplied by 10×10, that is, by 100. For example, 5734.0 is 100 times greater than 57.34. Shifting the decimal point 3 places to the right increases the number 1000 times. Shifting the decimal point 3 places to the left reduces the number 1000 times.

Summary

The multiplication of a number by 10 is expressed by a shift of the decimal point one place to the right, for example, $2.34 \times 10 = 23.4$. Division by 10 is expressed by a shift of the decimal point one place to the left, for example, $23.4 \div 10 = 2.34$.

Transforming Decimal Fractions into Ordinary Fractions

Any decimal fraction can also be expressed as an ordinary fraction. Let's take, for example, the decimal fraction 0.24. Transforming it into an ordinary fraction only requires interpreting what is written. The digit 2 represents tenths, and the 4 represents hundredths. Altogether we have 2 tenths and 4 hundredths. That is, $\frac{2}{10} + \frac{4}{100}$, or simply (after using a common denominator): $\frac{24}{100}$.

There is a simpler way to look at this. Move the decimal point in 0.24 two places to the right. This is the same as multiplying by 100. The result, 24, is thus 100 times greater than the original number. To return to the original number, the result must be *divided* by 100, yielding $24 \div 100$ or, as a fraction, $\frac{24}{100}$.

Transforming Ordinary Fractions into Decimal Fractions

We will now study the reverse — turning an ordinary fraction into a decimal fraction. This is very useful since calculations in decimal numbers are often simpler than in fractions. As an example, we will turn $\frac{3}{8}$ into a decimal fraction.

The principle is simple: Remember that fractions are division, that is, $\frac{3}{8} = 3 \div 8$. All we need to do is to divide.

Let's recall how this is done. Write:

$$\overset{0}{8\,)\,\overline{3}}$$

This is the first step in the calculation: 8 isn't contained even once in 3. Therefore, the result includes 0 ones, written above the 3, which is the ones digit of the dividend.

As may be recalled, the next stage in long division is changing the 3 for smaller units, to enable their division by 8. In this case, we will change the 3 ones into tenths: 3 is 30 tenths. A zero is added to the right of the 3, but to clarify that these are 30 tenths and not 30 ones, we will write a decimal point:

$$\overset{0}{8\,)\,\overline{3.0}}$$

Now divide the 30 tenths by 8. The result is 3, since 8 is contained in 30 three times. But remember: 3 is 30 *tenths*, therefore the result is 3 tenths.

This fact is marked by placing a decimal point in the result as well, above the dividend's decimal point:

$$\begin{array}{r} 0.3 \\ \hline 8 \overline{)\ 3.0} \end{array}$$

From here on, it is just like ordinary long division. The result, 3, is multiplied by the divisor, 8, resulting in 24. This means that 24 tenths were divided, and $30 - 24 = 6$ tenths remain undivided. It is written thus:

$$\begin{array}{r} 0.3 \\ \hline 8 \overline{)\ 3.0} \\ -2.4 \\ \hline 0.6 \end{array}$$

Now the 0.6, or 6 tenths, must be divided by 8. To do so, they are changed to 60 hundredths. We know that 60 contains 8 seven times, and therefore, 7 is written in the hundredths place of the result. However, there is a remainder: 7 times 8 isn't 60, it is 56. There are $60 - 56$, that is, 4 more hundredths to divide:

$$\begin{array}{r} 0.37 \\ \hline 8 \overline{)\ 3.0} \\ -2.4 \\ \hline 0.60 \\ -0.56 \\ \hline 0.04 \end{array}$$

The 4 hundredths are 40 thousandths (a point clarified by the addition of a 0 to the right of the 4). When divided by 8, the result is 5 thousandths, with no remainder. The final result is 0.375.

$$\begin{array}{r} 0.375 \\ \hline 8 \overline{)\ 3.0} \\ -2.4 \\ \hline 0.60 \\ -0.56 \\ \hline 0.040 \\ -0.040 \\ \hline 0.000 \end{array}$$

What we actually learned here is how to divide any pair of integers to result in a decimal fraction. There is only one problem:

Sometimes the algorithm just doesn't end. It may continue indefinitely. When dividing $1 \div 4$, for example, the result is 0.25, but when dividing $1 \div 3$, that is, when expressing $\frac{1}{3}$ as a decimal fraction, the result is $0.333\ldots$, the algorithm will add 3 with each step and will never end.

When does each case occur? The answer is given in the following section.

Which Fractions Can be Written as Finite Decimal Fractions?

The rule is that a fraction can be written as a finite decimal fraction only if its denominator contains the factors 2 and 5 alone, that is, it is divisible only by 2 and 5 and their powers. For example, 8 contains only the factor 2, since it equals $2 \times 2 \times 2$, and indeed $\frac{1}{8} = 0.125$. One hundred also contains the factors 2 and 5 alone, since it is equal to $2 \times 2 \times 5 \times 5$, and indeed $\frac{1}{100} = 0.01$.

To understand why this type of fraction can be written as a finite decimal fraction, let's examine the fraction $\frac{1}{8}$. The only factor in the number 8 is 2, since $8 = 2 \times 2 \times 2$. If so, $\frac{1}{8} = \frac{1}{2 \times 2 \times 2}$.

The secret is this: 2 can be multiplied by 5 to equal 10. Therefore, if the fraction is expanded by multiplying both the numerator and the denominator by $5 \times 5 \times 5$, the denominator will equal $10 \times 10 \times 10$, namely:

$$\frac{1}{8} = \frac{1}{2 \times 2 \times 2} = \frac{5 \times 5 \times 5}{10 \times 10 \times 10} = \frac{125}{1000}.$$

And this fraction can be written as a finite decimal fraction: 0.125. In general, if the factors of the denominator are 2 and 5, it can be multiplied to result in a power of 10. Therefore, the fraction can be expanded so that the denominator is a power of ten, yielding a finite decimal fraction.

We have demonstrated that fractions with a denominator containing only the factors 2 and 5 (each of which may appear several times) can be written as finite decimal fractions. Are there other fractions that share this property? The answer is no. Only fractions with a denominator containing the factors 2 and 5 alone can be written as a finite decimal fraction.

Why? Consider a decimal fraction such as 0.166. As an ordinary fraction it is $\frac{166}{1000}$. The denominator contains multiples of 10 only — for 1000 is a power of 10. Since $10 = 2 \times 5$, the factors of the denominator are 2 and 5 alone. For example, $\frac{3}{14}$ cannot be written as a finite decimal fraction, for $14 = 2 \times 7$, and the factor 7 prevents writing the fraction as a finite decimal fraction.

Repeating Decimal Fractions

An infinite decimal fraction is called repeating if from a certain point on its digits appear in a recurring order. For example: 0.512121212 ... is a repeating fraction, since the digit pair 12 repeats itself from the second digit after the decimal point and on.

The rule is that a repeating decimal fraction is always an ordinary fraction. We have already encountered one example: 0.333 ... is $\frac{1}{3}$. Let's prove this rule by demonstrating how a repeating fraction is transformed into an ordinary fraction. The proof is based on the odd fact already mentioned, that the number 0.999 ... equals 1. Let us use this fact to show that 0.333 ... $= \frac{1}{3}$. Write: 0.333 ... $= \frac{0.333...}{1}$. This is permitted, since dividing by 1 does not alter the number. Remembering that $1 = 0.999$ ..., the denominator can be replaced by 0.999.... Thus

$$0.333\ldots = \frac{0.333\ldots}{0.999\ldots}.$$

For each 3 in the numerator there is a 9 in the denominator. Therefore, the numerator is 3 times smaller than the denominator, and so the fraction equals $\frac{1}{3}$.

Here's another example: transforming the fraction 0.141414 ... into an ordinary fraction. Write:

$$0.141414\ldots = \frac{0.141414\ldots}{1} = \frac{0.141414\ldots}{0.999999\ldots}.$$

For each 14 in the numerator, there is a 99 in the denominator. Therefore, the fraction equals $\frac{14}{99}$.

Finally: How is 0.5121212 ... written as an ordinary fraction? I'll leave the reader to solve this one on his or her own, with a small hint. Write:

$$0.5121212 \ldots = 0.5 + 0.0121212 \ldots = 0.5 + \frac{1}{10} \times 0.121212 \ldots.$$

Adding and Subtracting Decimal Fractions

In the chapter *The Three Mathematical Ways of Economy* we mentioned a useful joke about mathematicians: "This is something we have already solved" (how to boil water). It is most useful when calculating with decimal fractions. For indeed, "this," that is, calculating the four arithmetical operations, is something we have already solved — in whole numbers.

There is nothing new to the addition and subtraction of decimal fractions. They are calculated in exactly the same way as are ordinary decimal numbers — vertically. The only thing to remember is to place the numbers so that the decimal points are one beneath the other. Example:

$$2.34$$
$$+\underline{15.8\ \ }$$

In this way of writing, the ones are beneath the ones, the tens beneath the tens, and the same is true to the right of the decimal point. The tenths are beneath the tenths, the hundredths beneath the hundredths, and so forth. This creates ready-made common denominators: Tenths are added to tenths, hundredths to hundredths.

What about the empty space beneath the 4? What does it represent? An empty space means 0. The 4 in the number 2.34 represents hundredths, and in the number 15.8 there are no hundredths. A 0 can be placed instead of the hundredths, 15.8 = 15.80 where zero hundredths are added, changing nothing.

How are Decimal Fractions Multiplied?

How is 2.3×0.75 calculated?

The answer: Ignore the decimal points, multiply, and then fix the problem created by ignoring the decimal points. In this case, 23 is multiplied by 75, resulting in 1725. Then, the result is repaired by dividing it by 1000, which means shifting the decimal point three places to the left to get 1.725.

Why did we divide by 1000 at the end? Because ignoring the decimal points means multiplying by 1000. Replacing 2.3 with 23 means shifting the decimal point one place to the right, or multiplying by 10, since $23 = 10 \times 2.3$. Similarly, replacing 0.75 with 75 means shifting the decimal point 2 places to the right, or multiplying by 100, since $75 = 100 \times 0.75$.

What happens when 23 is multiplied by 75? 23 is 10 times greater than the number in the exercise, and 75 is 100 times greater than the number in the exercise. Therefore, their product is 10 times 100, that is, 1000 times greater than the correct result of the exercise. To repair this, the result must be *divided* by 1000.

Here is another example: 0.2×3.333. Again, the decimal points are ignored, so that 2 is multiplied by 3333. The result is 6666. The decimal point in the multiplier was shifted one place to the right, changing the 0.2 to 2; the decimal point in the multiplicand was shifted 3 places to the right. Therefore, to repair the result, the decimal point needs to be shifted $1 + 3$, that is, 4 places to the left.

Since $6666 = 6666.0$, shifting the decimal point 4 places to the left results in 0.6666.

It's a good idea to make sure the result makes sense. 0.2 is a fifth; 3.333 is near 3. It makes sense that a fifth of a number near 3 is near 0.6.

Expanding Ratios of Decimal Fractions

Take a look at the exercise $6.33 \div 0.2$. It can also be written with a fraction line: $\frac{6.33}{0.2}$. Let's move the decimal point in both the numerator and the denominator one place to the right: $\frac{63.3}{2}$. What happened here? Both the numerator and the denominator were multiplied by 10. In other words, the fraction was expanded by 10, which doesn't alter its value.

When written in the language of division instead of fractions, what we have shown is that $6.33 \div 0.2 = 63.3 \div 2$. Notice, by the way, how much easier it is to estimate the size of $63.3 \div 2$ (which is about 30), than that of $6.33 \div 0.2$!

The rule is that in division the decimal point can be shifted the same number of times, in the same direction, in the dividend and the divisor, and the value of the quotient does not change.

How Are Decimal Fractions Divided?

Teaching division of decimal fractions should not be a high priority. Most of the principles were taught in long division and the multiplication of decimal fractions. But for those who do reach this point, here are the common calculation methods. There are three of these.

First Method: Ignoring the Decimal Point

One way to divide decimal fractions is similar to multiplying them: Ignore the decimal point and make up for it later.

For example, to calculate $33 \div 0.02$ first calculate $33 \div 2$, which is 16.5. However, this isn't the real result, of course, but only a hundredth of it: When the 0.02 was replaced with a 2, the divisor was increased 100 times, reducing the result 100 times. To rectify this, the result should be multiplied by 100. This is achieved by shifting the decimal point two places to the right: 16.5 becomes 1650, which is the correct answer.

Second Method: Expanding the Divisor and the Dividend into Integers

A second way to divide decimal fractions is by shifting the decimal point in the same manner in both divisor and dividend, until they both become whole numbers.

Take, for example, the exercise $0.123 \div 0.4$. Shifting the point 3 places to the right in the dividend and in the divisor makes them both integers while not changing the quotient, since both divisor and dividend are thus multiplied by 1000. After the shift the exercise is $123 \div 400$, or expressed as a fraction: $\frac{123}{400}$. It remains to transform this fraction into a decimal fraction, but we already know from the previous chapter how to do this.

Third Method: Shifting the Decimal Point until the Denominator becomes an Integer

The most popular method of all is probably the third one, which is also based on expansion but doesn't insist on both the divisor and the dividend becoming integers: It is satisfied with turning the divisor into an integer. For instance, in the exercise $0.123 \div 0.4$, the decimal points in the divisor and the dividend are shifted one place to the right resulting in $1.23 \div 4$. Then, the calculation is performed using ordinary long division.

Percentages — a Universal Language for Fractions

The Esperanto of Fractions

I usually begin teaching percentages with the story of the Tower of Babel. I ask one of the children to repeat the story, especially its end, about how God confounded people's languages. We discuss the advantages of having one common language for all people.

An ophthalmologist from Warsaw named Zamenhoff (1859–1917) tried to make the dream of a universal language come true. He invented a new language, called Esperanto, and hoped that it would catch on throughout the world. The fact that he was somewhat successful, and that currently more than a million people speak Esperanto, can be attributed to the simplicity and logic of the language, which do not come at the expense of its richness.

In fractions, the common language is a common denominator. Imagine how convenient it would be if all fractions had the same denominator! How simple it would then be to add and subtract them, and how easy it would be to compare them!

The world of fractions had its own Zamenhoff, someone who tried to find a universal common denominator. This took place in Italy, in the sixteenth or seventeenth century. The common denominator selected was 100. The fraction $\frac{1}{100}$ was, therefore, provided with a special name: "percent." The term originates from Latin, meaning "for each hundred."

1% is simply a hundredth. 50% are 50 hundredths, namely, a half. 100% are 100 hundredths, namely, a whole. 0.1% is a tenth of a hundredth, that is, a thousandth.

This tool enables us, among other things, to easily compare fractions. For example, saying that the unemployment rate in Israel is 10.3% and in Greece it is 11.5% allows us to realize straight away where the unemployment rate is higher.

Lo and behold! The wolf shall dwell with the lamb! Why didn't anyone think of this before? Isn't there some sort of catch?

A Number to Fit Man's Measurements

Well, the truth is that the invention of percentages isn't that ingenious. It doesn't contain any new mathematical idea, only practical advantages. Percentages aren't that different from decimal fractions. For example, 24%

just means 24 hundredths, that is, $\frac{24}{100}$ or $24 \div 100$, which is 0.24. Anything that can be written as percentages can also be written as decimal fractions. If so, what's so useful about the invention of percentages?

The advantage of percentages lies in the fact that, for many practical uses, they constitute a measurement tool with just the right degree of fineness. The choice of the universal common denominator wasn't entirely arbitrary. To facilitate calculations such a denominator must be a power of 10 — the basic fraction must be 1, or $\frac{1}{10}$, or $\frac{1}{100}$, or $\frac{1}{1000}$, and so forth. It turns out that of all of these, dividing into hundredths is most suitable for the measurements people use. In many cases, it has exactly the required degree of fineness. For example, the results of an election usually require more accuracy than that provided by tenths, but hundredths are sufficient. If Candidate A received 44% of the votes, and Candidate B 41%, in the language of tenths both would be rounded off to $\frac{4}{10}$, and we wouldn't know who won. The accuracy provided by percentages is also adequate for purposes of income tax, or sales tax, which is usually defined as whole percents.

Percentages also suit people's measurements in that they are easy to calculate. Calculations in percentages involve two-digit numbers, which most people can handle.

How to Switch from Percentages to Fractions, and Vice Versa?

How are 24% turned into a fraction? Nothing is easier. 24 percent are, according to the definition of "percent," 24 hundredths, that is, $\frac{24}{100}$; the number 24 is divided by 100. To make the transition from percentages to fractions, divide by 100.

What about the reverse direction? How are fractions turned into percentages? If the transition from percentages to fractions requires division by 100, then the opposite transition requires multiplication by 100. For example, to turn $\frac{1}{4}$ into a percentage, it must be multiplied by 100, that is, $100 \times \frac{1}{4} = 25$, meaning that $\frac{1}{4}$ is 25%. Let's examine this result: 25 percent is $\frac{25}{100}$, and after reduction, $\frac{1}{4}$.

The same can also be demonstrated directly. To express $\frac{1}{4}$ in percentages, we ask: How many percents does $\frac{1}{4}$ contain? In other words, how many hundredths does a quarter contain? To find out, for example, how many times the number 2 is contained in the number 6, we divide 6 by 2. Similarly, the number of hundredths contained in a quarter is $\frac{1}{4} \div \frac{1}{100}$. However, dividing by $\frac{1}{100}$ is the same as multiplying by 100. Therefore, the answer is $\frac{1}{4} \times 100$, as seen above.

Summary

To turn a fraction (or, in fact, any number) into a percentage, it is divided by 100. To turn a percentage into a fraction, it is multiplied by 100.

A 10% Addition to the Salary

The owner of a coat shop went on some errands, leaving the new salesman to watch over the store. When he returned, the salesman told him proudly that he sold the leather jacket for 100 dollars. What? cried the owner. The listed price was 1000 dollars! After a moment he calmed down. Oh well, we made a 100% profit on it anyhow.

Practice question: What would the owner's profit have been had the jacket been sold at the original price? Does the fact that the listed price is 10 times greater than the selling price mean that the profit is also 10 times greater? (The answer is "no," and I'll leave it to the reader to discover why.)

This is an example of practical problems involving percentages that are often encountered in everyday life. This section and the next will be dedicated to such problems. In fact, these are no different from fraction problems, but since in everyday life they usually appear as percentages, their proper place is in this context. The first problem is of a familiar kind:

Following a 10% raise, your salary is 1100 dollars. What was your salary before the raise?

It seems very simple! 10% was added? To return to the original salary, 10% should be subtracted! 10% of 1100 is 110. So your previous salary was 1100–110, that is, 990 dollars!

This is a common mistake. Let's examine it closely. Suppose the original salary was indeed 990 dollars. 10% of 990 is 99 dollars. 990 plus 99 is 1089, and not 1100!

The mistake is that the reverse of adding 10% to the original salary is subtracting 10% of the *original* salary, and what we did was to subtract 10% of the salary *after* the raise.

However, this doesn't lead us to a solution. How can 10% of the original salary be subtracted, if the original salary is unknown? After all, the original salary is precisely the amount sought!

The secret to the solution is this:

Adding 10% is the same as multiplying by 1.10.

10% of a number is a tenth of it. Adding 10% means adding a tenth. The number itself is, of course, $\frac{10}{10}$ of itself. After the addition of a tenth, there will be $\frac{11}{10}$. In other words, adding a tenth is the same as replacing the number with $\frac{11}{10}$ of itself, which means multiplying it by $\frac{11}{10}$, or multiplying the number by 1.1.

Now we can answer the above mentioned question. An addition of 10% to a salary is like multiplying the salary by 1.1. Therefore, to return to the original salary, the new salary must be divided by 1.1. If the multiplication resulted in 1100, the original salary was $1100 \div 1.1 = 11000 \div 11 = 1000$, that is, 1000 dollars.

Here is another example of this type:

The price of a garment with a 20% purchase tax added is 300 dollars. What was its price before tax?

Adding a 20% tax means multiplying by 1.2. To return to the original price, before tax, the price is divided by 1.2. The price before tax was:

$$300 \div 1.2 = 300 \div \frac{6}{5} = \frac{300 \times 5}{6} = 250.$$

E. Ratios

Ratios integrate in one subject much of the material taught in elementary school: division, fractions, and their connection to real-life situations. Ratio problems are also useful since reality often obeys rules of constant ratios between quantities called "direct proportion" or "proportionality."

The main tool used to handle ratio problems is "rate per one unit," that is, how many units of type A there are per one unit of type B. For instance, how many kilometers a car travels in one hour.

Proportionality

> *There are days when the greens Are seven times greener And the blue above is seventy times bluer.*
>
> Rachel, **Kineret**

Ratio Problems

What is "maturity"? I wish I knew ... "Mathematical maturity," however, is easier to define. It means internalization of the mathematical way of thinking. Someone who has just acquired a concept will grope for its correct usage. One who internalized it will assuredly know to which situations it applies. Mathematical maturity means confidence in handling mathematical arguments.

The final subject in elementary school arithmetic is ratios. Ratio here is related to proportionality; it is a relationship — how many times one quantity is greater than the other, rather than how large each quantity, which might be changeable, is in itself. No other subject is more appropriate as the grand finale to elementary school, making it a real test of mathematical maturity. It requires internalization of the concepts of division and ratio, and understanding of the connection between arithmetical operations and real life, that same connection we called "meaning."

Direct Proportion

Tree A is twice as tall as tree B. How many times is the shadow of A longer than the shadow of B?

Of course, the shadow is twice as long, too. This phenomenon is called "proportionality," or "direct proportion." It means a constant ratio between two sizes. For example, between the height of a tree and the length of its shadow.

The world is full of phenomena involving proportionality. If 3 times more people than originally planned go on a trip, they will need to take 3 times more provisions. They will also have 3 times more fingers and 3 times more ears.

Another example: Shapes and images are identified through direct proportion. The same object leaves a different impression on the retina of the eye when it is near or far. We know that it is the same object in both cases because the ratios between the various components are the same in both.

As usual, the first step is getting familiar with examples. Here are a few:

The ratio between the number of cars and the number of their wheels is always a quarter (not counting the spare wheels).

The ratio between the number of meters and the number of kilometers traveled is always a thousand.

The previous example can be generalized. A constant ratio applies to any transition between different units. One can show the children a ruler and demonstrate that the length in centimeters is always two and a half times (unfortunately, not exactly) the length in inches. One can also use this to explain the ratio of five to two, (in mathematical writing 5:2), where every 5 centimeters equal approximately two inches.

Want to bake a cake 1.5 times larger than the one in the recipe? Each of the ingredients needs to be increased by a factor of 1.5. Instead of 4 eggs, you'll need 6. Instead if 2 cups of flour, you will need 3.

I know a teacher who teaches direct proportion by examining the ratio between the children's heights and the length of their shoes. The children lie on the ground, and mark their height with a chalk. Then, they take "heel to toe" steps and mark how many times the shoe is contained in the child's height. This ratio is more or less constant: 6 (7, if they're barefoot). The taller the child, the bigger the size of his shoes. An adult should also be included in the experiment!

Direct proportion can be demonstrated through measurements of shadows. The heights of the children and the teacher are measured and compared with the lengths of their shadows. The ratios are then examined: If the teacher is $\frac{3}{2}$ times taller than the student, her shadow is also $\frac{3}{2}$ times longer.

Measuring the diameters and circumferences of various circles demonstrates that the ratio between the two is constant (the circumference is approximately 3.14 times greater than the diameter).

A teacher I am acquainted with conducts a "Gulliver Day" at school: Gulliver is 10 times smaller than we are (assuming we are the giants). What will his table look like? His plate? The students create cardboard furniture and tools for him. They measure the furniture and tools around them and divide their measurements by 10.

Two Ways to Express Direct Proportion

It is important to note that direct proportion between two quantities can be expressed in two different ways. To see this, look for example in the fact that there is direct proportion between the number of people and the number of their fingers. This can be expressed in two ways:

For any two groups of people, A and B, the ratio between the number of fingers and the number of people in each of the two groups is the same. (It is equal to 10).

Or:

The ratio between the numbers of fingers in both groups is equal to the ratio between the numbers of people in both groups. (For example, if Group A includes 3 times more people, then it also includes 3 times more fingers).

For those familiar with algebra, the equivalence between the two wordings is simple. Define a and b as the number of people in Groups A and B, respectively, and x and y as number of their fingers. The first version says that

$$\frac{x}{a} = \frac{y}{b},$$

while the second version says that

$$\frac{b}{a} = \frac{y}{x}.$$

The second formulation can be derived from the first, and vice versa, because both express the same equality of the products:

$$a \times y = b \times x.$$

The King's Road to the Solution of Ratio Problems: Rate per One Unit

Direct proportion involves four quantities. For example, the height of each of two trees and the length of each of their two shadows, or the numbers of people in two groups and the respective numbers of fingers. The problems concerning these quantities often have the same format: Three of the four quantities are given, and the question concerns the fourth. For instance: The height of one tree is 4 meters, and the length of its shadow is 6 meters. (An intermediate question — can you say anything about the time of day?). The height of another tree is 10 meters — how long is its shadow?

As in many cases, the secret to the solution of ratio problems lies in finding an intermediary stage, a step that when added to the ladder makes climbing it easier. In this case, the intermediary step is the *rate per one unit*.

There is one type of ratio problem where this principle is familiar to all: problems of motion, namely time-distance-speed. These aren't usually considered very difficult. Why? Because someone bothered to invent the right intermediary stop for us: measuring the speed in *kilometers per one hour*. For instance:

A car travels 120 kilometers in 2 hours. How far will it travel in 3 hours?

The secret lies in the calculation of speed. If a car travels 120 kilometers in 2 hours, its speed is 60 kilometers per hour. In 3 hours it will therefore travel 3 times 60, that is, 180 kilometers.

We are so used to this solution that we do not appreciate its wisdom. Quantities A and B are given (in this case, the distance measured in kilometers and the time measured in hours), and the question is: How many units of A are there per one unit of B? This is the "rate per *one* unit," or simply "rate."

Stage One: The Use of Rate

The solution of ratio problems consists of two stages. The first is calculating the rate, and the second is *using it*. Of the two, the second is simpler, since it requires multiplication, rather than division. Therefore, it is recommended to start teaching ratio problems with it. Here is an example:

A car travels 50 kilometers in an hour. How many kilometers will it travel in 3 hours?

The answer is, of course, 3 times 50, which equals 150 kilometers. This kind of example should be repeated over and over, *ad nauseam*: If a car travels 40 kilometers in an hour, how far will it travel in $3\frac{1}{2}$ hours? If a worker digs 3 pits in an hour, how many will he dig in 4 hours? If Danny receives an allowance of 10 dollars each week, how much will he receive in 5 weeks?

Stage Two: Calculation of Rate

As mentioned, when solving ratio problems the first stage is calculating the rate. When teaching, it should come second, since it requires division and is thus more difficult. To calculate the rate of Quantity A to Quantity B (that is, how many units of Quantity A there are per one unit of Quantity B), Quantity A is divided by Quantity B. For example:

A group of planters plant 90 trees in 3 hours. How many trees will they plant in 1 hour?

In one hour the group will plant 3 times less than in 3 hours, that is, $90 \div 3$, which is 30 trees. This stage should also be repeated again and again: A car travels 100 kilometers in 2 hours. How far will it go in 1 hour? A tree 2 meters high casts a 6 meter shadow. How long a shadow does a 1-meter-high tree cast?

Combining Both Stages

After both stages have been thoroughly internalized, proportionality problems become simple.

A worker paves 50 tiles in 2 hours. How many tiles will he pave in 5 hours?

As mentioned, the first stage is calculating the rate: Each hour, the worker paves $50 \div 2 = 25$ tiles. The second stage is using the rate: If he paves 25 tiles per hour, in 5 hours he will pave 5 times as much, that is, 125 tiles.

Here is another type of question:

A group of workers plants 90 trees in 3 hours. How long will it take them to plant 600 trees?

Again, the solution begins with calculation of the rate, namely the number of trees planted in 1 hour. Each hour, the group plants $90 \div 3 = 30$ trees. Planting 600 trees will take the number of times 30 must be multiplied to result in 600, that is, $600 \div 30$, which is 20 hours.

Another option: The required number of trees, 600, is $\frac{600}{90}$ times greater than the given number of trees, 90. Therefore, the time will also increase $\frac{600}{90}$ times, resulting in $3 \times \frac{600}{90} = 20$.

A third possibility is this: How long does it take to plant 1 tree? If 90 trees take 3 hours, 1 tree takes $\frac{3}{90} = \frac{1}{30}$ hours. Thus 600 trees will take $600 \times \frac{1}{30} = 20$ hours.

The Rule of Three

Leonardo of Pisa, also known as Fibonacci (1200–1260) made some profound contributions to mathematics. But one of his contributions to mathematical education was of doubtful value: a pattern for solving problems of direct proportion, called "The Rule of Three": Three given quantities are placed at the vertices of a triangle and some recipe is applied in order to find the fourth quantity.

I cannot count the number of adults who have told me that "The Rule of Three" still haunts their dreams. I have seen children (including my own son) who understood ratio problems until they were forced to study this pattern. When a pattern does not result from comprehension, it induces anxiety. After a month or two the pattern is forgotten and the anxiety remains.

More Ratio Problems

This chapter deviates from the material currently taught in elementary schools. When I was a child these subjects were taught in the sixth grade, and in some countries they still are. Their main significance is in their contribution to the internalization of the concept of ratio.

Inverse Ratio

If 3 workers pave a road in 10 days, how many days will it take 5 workers to pave it?

This problem concerns the ratio between two quantities: the number of workers and the number of days. But consider: What happens when there are more workers? Do they also require more days? Of course not — they require *less* days. If the number of workers is twice more, the number of days is twice *less*.

The two quantities are "inversely proportional." When one increases *times* a certain number, the other decreases *times* the same number.

The secret of inverse ratio problems also lies in manipulating one quantity to equal a single unit. (A customary name for this is "bringing to a unit.") For example, calculate how long it takes 1 worker to do the job. To accomplish what 3 workers accomplished, he needs 3 times more days, that is, 3 times 10, which are 30 days. The second stage of the solution: 5 workers need 5 times less days than 1 worker. That is: $30 \div 5$, which is 6 days.

Another option is to manipulate the second quantity, time, into a single unit: Calculate what would happen if there were one day instead of 10. This would require 10 times more workers, that is, 30 workers. If 30 workers pave a road in 1 day, then 5 workers require $30 \div 5$ times more time, namely 6 days.

Pool Problems

One tap fills a pool in half an hour. A second tap fills it in an hour. How much time will it take to fill the pool if both taps are open?

Many people remember "pool problems" with a shudder. Indeed, they contain an inherent difficulty. One teacher told me about how he had taught a private student pool filling problems for weeks. She returned in tears from the examination: There weren't any pool questions. There was a problem about a movie theater which fills in an hour when one door is open and in half an hour when the other door is open, and the question was how long it takes to fill the theater when both doors are open.

The Secret of Pool Problems: Inverse Rate

If everybody minded their own business, the world would go round a deal faster than it does.

The Duchess, from *Alice in Wonderland*

by Lewis Carroll

If teachers told their students the secret of pool problems, a lot of suffering would be spared. The secret lies in the idea of the inverse rate.

If a rate is the number of units of Quantity A per each unit of Quantity B, then the inverse rate is defined as the number of units of Quantity B per each unit of Quantity A.

For example: Speed is the number of kilometers per one hour (Quantity A here is the distance measured in kilometers, Quantity B is the time measured in hours). The inverse rate is the number of hours per one kilometer. For instance, a car travels 40 kilometers in 1 hour. How long will it take it to travel 1 kilometer? The answer is $\frac{1}{40}$ of the time it takes to travel 40 km, that is, $\frac{1}{40}$ of an hour. In general, inverse rate is one over the original rate.

In pool problems, the idea is that instead of asking how many hours it takes to fill one pool, you ask how many pools are filled in 1 hour. For instance, a tap fills one pool in 3 hours. How many pools will it fill in 1 hour?

The answer is, of course, $\frac{1}{3}$. The inverse rate is $\frac{1}{3}$ of a pool per hour — one over the original rate, which was 3 hours per pool.

Or, if we return to the tap in the problem at the beginning of this section, it filled a pool in $\frac{1}{2}$ an hour. How many pools will it fill in an hour? 1 over a $\frac{1}{2}$, that is, 2 pools. (This is also obvious — in one hour it will accomplish twice as much as in $\frac{1}{2}$ an hour.)

Why is it a good idea to turn to inverse rate? Because when discussing pools per hour, it is easy to see what both taps do together. If one tap fills 2 pools in an hour and the other fills 3 pools in an hour, together they fill $3 + 2$, which are 5 pools in an hour. This is simply an addition problem.

In fact, when teaching pool problems it is a good idea to begin with this type of problem before presenting the original type. One tap fills half a pool in an hour and the second tap fills a third of a pool — how many pools will they fill altogether in an hour? One tap fills 2 pools in an hour and the other

4 — how many will they fill together in an hour? And so on, *ad nauseam*, until things become self-evident.

Now we know how to solve the original problem: Turn to inverse rate and solve the simple kind of problem, which requires no more than addition. As recalled, the first tap fills a pool in half an hour, and therefore, it will fill twice as much in an hour — 2 pools. The second tap fills one pool in an hour. This is a given, there is no need to calculate the inverse rate. How many do both taps fill altogether in an hour? This is the easy kind of problem we discussed: Altogether, they fill $1 + 2 = 3$ pools in an hour.

Only one stage remains: If both taps fill 3 pools in an hour, how much time is needed to fill 1 pool? Of course, one third of this time, namely $\frac{1}{3}$ of an hour, and this is the answer to the question. The final transition can be deduced from the rule of inverse rate: The common rate of the taps is 3 pools per hour, therefore, the inverse rate is $\frac{1}{3}$ of an hour per pool.

Why are Pool Problems Difficult?

The reason is the need to add a factor that didn't exist in the original problem. Actually, two factors. First, the correct unit must be found. We are used to thinking of measuring water in terms of liters and cubic meters. Here, the correct unit is "one pool." The second new idea is inverse rate, asking how many pools a tap fills in one hour (rate per one unit of time) instead of how many hours it takes to fill 1 pool. This isn't entirely intuitive, especially since it is difficult to think of a tap filling 2 pools in an hour when there is actually only 1 pool! How will it fill 2 pools? Well, if it had been provided with another pool, it would have filled them both.

The Same with Cars

So as not to give the impression that these principles work only with pools (or movie theaters, for that matter), here's another example, this time with speed:

A car travels from Tel Aviv to Haifa in three quarters of an hour. A second car travels the same distance in an hour. How much time will it take them to meet if both leave at the same time, one from Tel Aviv to Haifa and the other from Haifa to Tel Aviv?

Again, it is important to first figure out the proper unit. In this case, the unit is "the distance from Tel Aviv to Haifa." The question is: How long does it take both cars to travel this unit together?

Now the rate must be inverted: Ask not how many hours it takes a car to travel the distance between Haifa and Tel Aviv, but how many times it will travel that distance in one hour. The first car travels the distance in $\frac{3}{4}$ of an hour. Therefore, in one hour it will travel the distance $\frac{4}{3}$ times. Like a tap that fills $\frac{4}{3}$ pools. The second car travels the distance once in one hour. Therefore, in one hour they will travel the distance

$$\frac{4}{3} + 1 = \frac{4}{3} + \frac{3}{3} = \frac{7}{3}$$

times together. Now we'll use the principle of inverse rate once again. In $\frac{3}{7}$ hours, they will travel the distance once altogether. This is the time it will take them to meet. The answer is, therefore, $\frac{3}{7}$ of an hour.

Mixture Problems

In a cattle food mixture there are 3 kilograms of millet and 4 kilograms of hay for every 2 kilograms of grass. A cow eats 18 kilograms of food mixture per day. How many kilograms of millet does it eat?

This type of problem is called "mixture problems." The ratios between the various quantities joined together to form a large quantity are given. What is each part if the whole is given?

According to the problem above, of each $2 + 3 + 4$ kilograms that the cow eats, 3 kilograms are millet. But $2 + 3 + 4 = 9$. Therefore, 3 of every 9 kilograms are millet, that is, $\frac{3}{9}$, which is $\frac{1}{3}$. Of the 18 kilograms the cow eats, $\frac{1}{3}$ is $\frac{18}{3} = 6$. If so, a cow eats 6 kilograms of millet.

For each 2 lines of a computer program that Joe wrote, Oliver wrote 3 lines. Altogether, they wrote 550 lines. How many lines did Joe write?

Of each $3 + 2$ lines, that is, 5 lines, that both wrote, 2 are Joe's. That is $\frac{2}{5}$. Two fifths of 550 is

$$\frac{2}{5} \times 550 = \frac{2 \times 550}{5} = \frac{2 \times 110}{1} = 220.$$

Therefore, Joe wrote 220 lines.

A Chicken and a Half

We'll end with a well-known riddle:

A chicken and a half lay an egg and a half in a day and a half. How many eggs does a chicken lay in a day?

If a chicken and a half lay an egg and a half in a given period of time, then one chicken lays one egg in the same period of time. In this case, the period of time is a day and a half. Therefore, one chicken lays one egg in a day and a half. In one day, it will lay

$$1 \div 1\frac{1}{2} = 1 \div \frac{3}{2} = \frac{2}{3} \text{ of an egg.}$$

Afterword

Of my lectures at university, some are better and some less so, but when over, they are usually forgotten. In elementary school my experience is entirely different. I come home either deeply upset because of a lesson that didn't work, or elated after lessons that were, at least as far as I am concerned, wonderful. One reason may be that the children's reactions are more direct than those of university students, and you know exactly where you went wrong and where you succeeded. But I think the real reason is that in a good lesson you actually touch the children. This happens primarily on an intellectual, not emotional, level. You cause the children to directly experiment with some principle, and your elation is a reflection of their response.

Throughout the book, I tried to convey this experience to the best of my ability. One book cannot cover the entire theory and practice of teaching arithmetic in elementary school. It is impossible to summarize six years of schooling in a book. Each lesson can give birth to a chapter of its own, each lesson requires many creative ideas with which to involve the children, and in each lesson more concepts are taught than the teacher or the children are aware of. It is only possible to describe the main framework of the material and to convey a few principles. This is what I tried to do. One message I attempted to impart is that the way to share mathematical ideas with the children is by direct experience, free of sophistication, based on respect for the complexity of elementary mathematics.

One cannot expect that, after reading the book, he will know and understand all subjects it touches, nor can all parents expect to help their children with every question. Lots of practice, exercising and experimenting are needed, on the side of the parent just as on the side of the child. I hope I succeeded in whetting the appetite of the reader for learning more.

Appendix:

Turning Points in the History of Modern Mathematical Education

Academization

Communism is a wonderful theory. There is only one problem — it can be implemented.

The last 50 years have witnessed much turbulence in the world of education. Few people in the general public are aware of the new developments, although they have had a far-reaching influence on our children's, and therefore our own, lives. The way our children spend the most substantial part of their day determines not only the quality of their lives, but also the nature of the society in which we will all live in not so many years.

In order to help your child, it is important to be familiar with the educational theories that shaped the way he or she is taught. One should know why the textbooks are arranged in a certain way, and according to which principles the classes are taught. This chapter describes some of the new educational trends.

The turning point occurred in the 1950s and 1960s. Following World War II, huge budgets were allocated to education in the West, and a roadmap was required. The job of supplying one was assigned to the universities' departments of education. "We are here," they have been announcing since then. With the passion and faith of revolutionaries, they outlined not only teaching methods, but often content decisions as well.

One major change was in the training of teachers, which was transferred from those who were basically teachers themselves to the hands of academicians. An important rule is that a teacher teaches what he knows and not what the student needs. A lecturer at a seminar, educated in higher mathematics, will convince himself that what the teachers need is higher mathematics. A lecturer who uses statistics for his educational research will believe that elementary school teachers need to know statistics. All these are, of course, at the expense of the basic mathematics taught in elementary school, the knowledge of which the teachers need so desperately. Concurrently, other developments occurred.

Educational researchers suddenly realized the enormous power they were given in the form of freedom to make far-reaching changes. In the decades since then, they have initiated and implemented several educational revolutions. The problem is that in education innovative ideas can be implemented within a short period of time, even if they haven't been thoroughly examined. The feedback system isn't efficient either: Students have no say in the methods of experiments upon them, and those outcomes that can be measured emerge too late and can always be attributed to many factors.

This doesn't concern only mathematical education, of course. Innovative theories were introduced into the teaching of reading and caused major strife in the United States. (In Israel their introduction stirred up a much smaller public storm.) Perhaps the most significant of all were two revolutions concerning the basic mode of teaching. One was the "personalized teaching," the other concerns class seating arrangements.

"Personalized teaching" means that every student advances at his own pace, according to his own capabilities. The result is that each student studies a different subject, implying that the teacher cannot teach the entire class. The mode of teaching switches to private tutoring, and one of the most powerful tools of teaching, the common discussion of the entire class, is lost.

The second, closely related change, was in classroom seating arrangements. The tradition of students sitting face-forward towards the teacher was replaced with sitting in groups, around tables, where half the children sit with their backs or profiles to the teacher. The seating arrangement has a decisive influence on the nature of the lesson, and on the teacher-student relationship. An aspect that was not given proper consideration was what this does to the teacher: Talking to student's backs can be rather unsettling. And, of course, this mode of seating entirely changes the teacher's status in the class.

However, in this appendix we will concentrate on mathematical education. The upheavals there began in the 1950s, and needless to say, they originated in the United States.

The "New Mathematics" and the Structural Approach

Bring forth the old because of the new.

Leviticus 26:10

In 1957, the Russians launched the first satellite, the Sputnik, into space. These were the years of the Cold War, and panic gripped America — the Russians were ahead of them in science. Within a short period of time, educationalists and mathematicians gathered to create a new curriculum that would turn children into little scientists. "There is no need to start at the beginning," wrote Sargent Shriver, brother-in-law of President Kennedy and head of the Peace Corps, in the introduction to a book that explained the program. "The children can begin from where the researchers are at." The idea was to teach the children abstract mathematics at an early age. This was called "The New Math."

Within a few years, the level of mathematical knowledge of American students hit rock bottom. Tom Lehrer, the famous mathematician — poet — composer — singer, wrote a song about children who know that $3 + 2 = 2 + 3$ but don't know what the sum is. In 1973, Morris Kline published *Why Johnny Can't Add* (following in the footsteps of Rudolf Flesch's book, from the 1950s, *Why Johnny Can't Read*, that discussed the effects of the "global reading" method). The grandiose ideas were completely abandoned in the 1970s. However, they didn't disappear: They were exported to less developed countries, through students who had studied for higher degrees in the Untied States and returned to their homelands with news of the revolution.

Probably nowhere was the effect as sweeping as in Israel. At the end of the 1970s, with the aid of a generous Rothschild donation (good intentions of donors combined with progressive ideas of educationalists are a universal recipe for disaster), a very strange system of books took over the Israeli mathematical education scene. The books were based on the New Math spirit, but took its ideas much further. The basic idea was that of indirect learning: Everything was taught via models, mostly invented by the creators of the system. Counting objects was discouraged, and using fingers for counting and calculating was forbidden. Instead, plastic sticks called Cuisenaire rods (after their inventor) were used as models for arithmetical operation.

The Advent of the Singapore Textbooks

Israel is also one place where a fight of a few mathematicians and teachers (I being one of them) led a determined struggle, and won. Within three years from the beginning of the struggle, the strange "structural" books were forbidden for use in schools. A hyper-constructivist (see the next section for the definition of the word) curriculum that was going to go into effect was replaced by a sane, demanding (unfortunately, sometimes over-demanding) and conservative curriculum.

An organization of which I am a member, the Israeli Foundation for Mathematical Achievement for All, is now working in over 10% of the schools in Israel, with the number rising rapidly from year to year. In our schools we are trying to implement principles similar to those described in this book, with apparent success. In the national tests held in 2006, all the schools studying with us were above the national average, and whereas the average in all schools in the country dropped with respect to the previous year's tests, all our schools improved their achievements.

Our foundation uses as textbooks translations of the Primary Math books which were used for many years in Singapore (they are now replaced in their country of origin, but by books which follow the same spirit and same outline). The spirit of these books, which emphasizes direct experience, stresses the meaning of the operations, encouraging the children to invent their own arithmetical stories, and systematic development of the topics, suits precisely the ideas of the people working in our foundation. It also slowly oozes to other systems of books used in the country.

Investigation and the Math Wars

In the United States, the storm over the New Math subsided toward the end of the 1970s. But the dust had hardly settled before a new revolution loomed on the horizon. Its name was "investigation," or "constructivism," signifying that the child should construct the knowledge on his own. No longer, said the proponents of this approach, should the child be a receptacle into which knowledge is poured. He should find things out by investigating the world. The teacher's role is that of a facilitator and supporter in the process of discovery.

This approach was implemented in the teaching of all areas of knowledge, but its effect was most significant in mathematics because of its highly layered structure of concepts. Rigor was no longer assumed to be necessary, resulting in what the opponents of the method named "fuzzy math." For example, fractions were taught through cutting sections of circles, and as critics observed, in this representation children are easily led to believe that $\frac{1}{3} + \frac{1}{4} = \frac{1}{2}$.

In 1989 the National Council of Teachers of Mathematics assembled the principles of the revolution in a book *The Standards*. The book calls for redefining the goals of mathematical education. These are not knowledge, so says the book, or calculation skills. The material is not the essence. The goal is to achieve investigational abilities and deeper insights, for example, knowing how to make assumptions or gather and process data. The words were exalting: creativity, placing the student in the center, teaching as a joint experience for teacher and student, self-discovery.

The first U.S. state to adopt the new approach was California. To the distress of Californian educators, the results were far from brilliant. Within five years, California dropped to the 48th place among the American states in comparative mathematical tests. The percentage of students requiring preparatory courses in mathematics upon entering the university increased two and a half times. Managers in California hi-tech companies realized that they had no local candidates to fill available jobs.

When the reports of failure continued to accumulate, a surprising thing happened: The exact science people awakened from their slumber. Many experienced the problem through their own children, who were guinea pigs of the system. Parents, mathematicians and scientists united and went to war. This was the starting point of the "math wars," which have been stirring the American educational system for nearly a decade. In a petition to the Secretary of Education, published in the *Washington Post* in 1998 as a paid advertisement, 220 well-known mathematicians and scientists called on the Secretary to renounce the new approach.

In California, where the revolution began, those opposed to the investigation approach prevailed. In 1997, mathematicians were appointed to write a new curriculum, and in 1998 it was introduced by law into all schools. The curriculum is arranged according to the traditional format: It determines what the children need to know, and not how to teach. The result, as reflected by the tests, worked wonders within a short period of time.

Systematization vs. Randomness

When Shalom Aleichem returned from a visit to Switzerland, he was asked if it is really as beautiful as people say. "The view isn't bad," he replied. "Too bad it's hidden by the mountains."

Anyone who has ever set foot in elementary school knows that the image of the teacher as a funnel and the students as receptacles is entirely preposterous. It is impossible to lecture in elementary school. The children won't listen even for one moment. Teaching children must be interactive, through experimentation and discussion. If so, what is special about the investigational approach? The secret is in systematic vs. random teaching. "Investigation" means forgoing systematization. Knowledge is not established heel to toe, with the teacher's guidance, or that of textbooks. Instead, random activities are pursued through which the child is supposed to discover mathematical structure on his own.

I believe that the true origin of this approach is in the misunderstanding of the depth of elementary school mathematics. In this sense, it is no different from high school or even higher mathematics. It is just that its principles are finer and less discernible. Just as we don't expect students to discover the principles of university mathematics without guidance, we cannot expect them to do so in elementary mathematics.

Another idea behind the investigational approach is the discovery of the beauty of mathematics. Simple arithmetical operations, according to this approach, are boring. They are a hurdle that must be overcome on the way to the real mathematics. The beauty of mathematics lies in creative activities.

As mentioned in the introduction, I also started out with such an approach, and I am no longer proud to say so. I learned that dessert cannot replace the main course, nor can mathematical diversions be taught to those who do not understand the foundations. Furthermore, the four arithmetical operations and the decimal system are not mountains hiding the view. They are the view, and the beauty lies within them.

Printed in the United States
By Bookmasters

Printed in the United States
By Bookmasters